About the Book

Plant pathology (also phytopathology) is the scientific study of plant diseases caused by pathogens (infectious organisms) and environmental conditions (physiological factors).[1] Organisms that cause infectious disease include fungi, oomycetes, bacteria, viruses, viroids, virus-like organisms, phytoplasmas, protozoa, nematodes and parasitic plants. Not included are ectoparasites like insects, mites, vertebrate, or other pests that affect plant health by consumption of plant tissues. Plant pathology also involves the study of pathogen identification, disease etiology, disease cycles, economic impact, plant disease epidemiology, plant disease resistance, how plant diseases affect humans and animals, pathosystem genetics, and management of plant diseases.

About the Author

Dr. Sunil Tiwari, born in 16 July 1973, has taught Agriculture science and Soil science for more than 32 years. He has done his B.Sc, M.Sc & PhD (Agriculture Science) from C.C.S. University, Meerut. He is know for his contributions in the field and science of agriculture. He belongs to the Horticulture Science and Soil and Water Conservation Society.He is also an Editorial Board Member and reviewer of several high impact International Journals.

Plant Pathology and Microbiology

Dr. Sunil Tiwari

AMIGA PRESS INC.
DELHI (INDIA)

AMIGA PRESS INC.
H.O.: 28-A, Indra Park,
Chander Nagar, Delhi-110051 (India)
Tel.: 011-22504028, +91-98736-57000
E-mail: amigapress@gmail.com
amigaeditorial@gmail.com
amigapub@gmail.com
info@amigapress.co.in
Website : www.amigapress.co.in

This edition is published by Amiga Press Inc.

ISBN: 978-93-87295-30-8

First Published 2019

Printed and bound by Replika Press Pvt. Ltd.

Preface

A microorganism is a unicellular life form that cannot be seen with the naked eye. There are more microorganisms on and inside our body than there are cells that make up our entire body. Microorganisms are also referred to by their shorter name - microbes. Microbiology is a science that studies microorganisms. This chapter discusses the main groups of unicellular organisms. It also looks at the different branches of microbiology.

The study of infectious disease is another important branch of microbiology. At the beginning of the 20th century infectious diseases, caused by microbial pathogens, were the major cause of death. Large numbers of children and the elderly succumbed to diseases such as tuberculosis, diphtheria and pneumonia. At this time microbiologists had little idea about how diseases were spread, or how they could be controlled, so epidemics flourished. The "Spanish" influenza pandemic of 1918–1919, caused ~50 million deaths worldwide - more than the total number of deaths recorded in World War One. Diarrhoeal disease was also common since people regularly ate contaminated food and drank contaminated water.

Plant disease, an impairment of the normal state of a plant that interrupts or modifies its vital functions.

Plant pathology (also phytopathology) is the scientific study of plant diseases caused by pathogens (infectious organisms) and environmental conditions (physiological factors).[1] Organisms that cause infectious disease include fungi, oomycetes, bacteria, viruses, viroids, virus-like organisms, phytoplasmas, protozoa, nematodes and parasitic plants. Not included are ectoparasites like insects, mites, vertebrate, or other pests that affect plant health by consumption of plant tissues. Plant pathology also involves the study of pathogen identification, disease etiology, disease cycles, economic impact, plant disease epidemiology, plant disease resistance, how plant diseases affect humans and animals, pathosystem genetics, and management of plant diseases.

The book Plant Pathology and Microbiology as taught in different courses in various universities are a source of information and mixture of all microbiological activities.

-Author

Contents

Preface *v*

1. History of Microbiology **1**

The History of Microbiology; Importance of Microscopes; Microbiological History; Spontaneous Generation; The Spectrum of Microbiology; Classification Schemes; Prokaryotes and Eukaryotes; The Five Kingdoms; Brief Descriptions of Microorganisms; Nomenclature of Microorganisms; Chemical Principles of Microbiology; Elements and Atoms; Molecules; The Immune System; Antigens; Cells of the Immune System; Antibody-Mediated (Humoural) Immunity (AMI); Cell-Mediated Immunity (CMI); Detecting Antibodies with LaboratoryTests; Disorders of the Immune System; Hypersensitivity Reactions; Immunodeficiency Diseases.

2. Identification of Microorganisms **34**

What is a Microorganism?; Land Microbes; Airborne Microbes; How are Microorganisms Identified; How do Microorganisms Reproduce; What do Microorganisms do; History; Evolution; Different Types of Microorganisms; Micro-Organisms Causing Disease; Why are Microorganisms such Powerful Pathogens; Protozoa; Microscopic Fungi; Classification and Structure; Chemical Requirements; Physical Requirements; Microbial Reproduction and Growth; Reproduction Patterns; The Growth Curve; Microbial Measurements; Pathogenicity of Microorganisms; Introduction; Communicable Diseases; Infection Transmission; Types of Pathogens; Epidemiology; History of Epidemics; Prevention of Infections; Bioterrorism; Beneficial Microorganisms.

3. **History of Plant Pathology** 79

Introduction; History of Plant Pathology; Defence against Plant Diseases; Objectives of Plant Pathology; Scope of Plant Pathology; Concept of Plant Disease; Causes of Plant Diseases; Classification of Plant Disease; Parasitism and Pathogenesis; Effect of Pathogen on the Plants; Symptoms of Plant Diseases; Development of Epidemics; Empirical Studies of Selection in Plant Pathology; Greenhouse Experiments to Measure Pathogen Fitness; Postharvest Plant Pathology; Preharvest Factors that Influence Postharvest Pathology; Postharvest Diseases of Fruits; Postharvest Diseases of Vegetables; Growth Regulators in Plant Pathology; Population Genetics and Phylogenetics in Plant Pathology.

4. **Nature & Importance of Plant Disease** 115

Principles of Plant Disease Management; Diagnosis of Plant Diseases; Nature and Importance of Plant Diseases; Disease Development and Transmission; Classification of Plant Diseases by Causal Agent; Principles of Disease Control; Bacterial Diseases of Beans; Pathogen Survival.

5. **Classification of Microbes** 144

What are Microbes ?; Decomposers; Symbionts; Disease Germs; Living with Microbes; Transmission; Symptoms; Diagnosis; Treatment; Classification of Microbes; Important Microbes; Bacteria; Fungi; Protozoa; Human Effected by Microbes; Benifits of Microbes; Environment Effected by Microbes; Microbes in Groundwater; Commonly Occurring Microbes; Pathogen Movement and Persistence; Protecting Public Health; Drinking Water Treatment; Surface Water; Other Treatment Technologies; Streams; Lakes; Wetlands; Algal Blooms in Fresh Water; Biodiversity; Wells and Well Drilling; Constructing a Well; Microbes Around us; Microbes are Helpful; Streak Plates; Microbial Enzymes; Microbial Rennets; Lactase; Proteinases and Peptidases; Proteolytic Enzymes; Production of Antibiotics; Preparation of Acrylamide; Identifying Useful Antibiotics; Types of Antibiotics.

Index 209

Chapter - 1

History of Microbiology

Microbiology is the study of microorganisms, which are microscopic, unicellular, and cell-cluster organisms. This includes eukaryotes such as fungi and protists, and prokaryotes. Viruses and prions, though not strictly classed as living organisms, are also studied. Microbiology typically includes the study of the immune system, or Immunology. Generally, immune systems interact with pathogenic microbes; these two disciplines often intersect which is why many colleges offer a paired degree such as "Microbiology and Immunology". Microbiology is a broad term which includes virology, mycology, parasitology, bacteriology and other branches.

A microbiologist is a specialist in microbiology and these other topics. Microbiology is researched actively, and the field is advancing continually. It is estimated only about one per cent of all of the microbe species on Earth have been studied.

Although microbes were directly observed over three hundred years ago, the field of microbiology can be said to be in its infancy relative to older biological disciplines such as zoology and botany.

THE HISTORY OF MICROBIOLOGY

A look at the history of microbiology will help you to understand the contributions of those who have come before. This perspective will hopefully give you an appreciation of their efforts and put the body of knowledge we will examine in the context of history. Keep in mind that microbiology is a relatively young science. It was only 130 years ago that it became possible to seriously study microorganisms in the laboratory, with most of our understanding of microbes coming in the last 60 years.

The history of microbiology, like all human history, is not a catalogue of linear progress, but is more of an interweaving of the careers of bright individuals and their insights.

Each new discovery relied on previous ones and in turn spawned further enquiry. A web of interdependent concepts evolved over time through the work of scientists in many related disciplines and nations. Often the research of one individual impacted the efforts of another studying a completely different problem. Keep this in mind as you look at this history.

Below we present several journeys through this web, mentioning some individuals who were particularly important in the progression. This history reflects our view of important events of the past, but is by no means comprehensive.

We will first look at the development of the techniques for handling microorganisms, since everything else in microbiology depends upon these procedures. Next, we will examine how these techniques helped to settle an old debate, the question of spontaneous generation. Then, we will look at the history of infectious disease. The science of microbiology had its most significant early impact on human health, uncovering the cause of the major killers of the day, and then methods to treat them. As microbiology matured, scientists began to look at what non-pathogenic microbes were doing in the environment and we will look a bit at the history of general microbiology. Finally, the chapter will end with an examination of the events that lead to the understanding of life at the molecular level and the profound impact this has had on microbiology and on society in general.

IMPORTANCE OF MICROSCOPES

It took the microscope to expose their tiny world and that instrument has been linked to microbiology ever since. In 1664, Robert Hooke devised a compound microscope and used it to observe fleas, sponges, bird feathers, plants and molds, among other items. His work was published in Micrographia and became a popular and widely read book at the time. Several years later Anton van Leeuwenhoek, a fabric merchant and amateur scientist (or "natural philosopher" as such people called themselves), became very adept at grinding glass lenses to make telescopes and microscopes. While crude by modern standards, his were a technical marvel for the time, able to magnify samples greater than 200-fold.

They also produced clearer images than the compound microscopes of the time. By peering through his microscope, Leeuwenhoek observed tiny organisms or "wee animacules" as he called them.. He spent months looking at every kind of sample he could find and eventually submitted his observations in a letter to the Royal Society of London, causing a sensation. Hooke was asked to confirm the findings of Leeuwenhoek and his affirmative assessment garnered them wide acceptance. Surprisingly the work of these two scientists was not followed up for almost 200 years. Human societies had neither the technical prowess nor the inclination to develop the science of microorganisms. It was not until the rise of the industrial revolution that governments and people dedicated the financial and physical resources to understand these small inhabitants of our world. With the development of better microscopes in the 19th Century, scientists returned to an examination of microorganisms. After finishing his education, Ferdinand Julius Cohn was able to convince his father to lay down the large sum necessary to purchase a microscope for him, one better than that available at the University in Breslau, then part of Germany.

He used it to carefully examine the world of the microbe and made many observations of eukaryotic microorganisms and bacteria. His landmark papers on the cycling of elements in nature was published in Ueber Bakterien in 1872 and a microbial classification scheme including descriptions of Bacillus were published in the first volume of a journal he founded, Beitraege zur Biologie der Planzen. Cohn's work with microscopes popularized their use in microbiology. This and his other work inspired many other scientists to examine microbes. Cohn's encouragement of Robert Koch, a German physician by training, began the field of medical microbiology.

MICROBIOLOGICAL HISTORY

Ancient man recognized many of the factors involved in disease. Early civilizations on crete, India, Pakistan and scotland invented toilets and sewers; lavatories, dating around 2800 BC, have been found on the orkney islands and in homes in Pakistan about the same time. One archaeologist has stated that "the high quality of the sanitary arrangements could well be envied in many parts of the world today". In Rome, 315 AD, the public lavatories were places where people routinely socialized and conducted business. Ten to twenty people could be seated around a room, with their wastes being washed away by flowing water; it must have been difficult to "stand on your dignity" under such circumstances.

The Chinese used toilet paper as early as ad 589. In Europe moss, hay and straw were used for the same purpose. I can personally attest to the use as late as 1962 of "slick magazines" as toilet paper in certain European camp grounds. The first cities to use water pipes (of clay) were in the Indus valley of Pakistan around 2700 BC. Metal water pipes were used in Egypt (2450 BC) and the palace of knossos on crete around 2000 BC had clay pipes.

Rome built elaborate aqueducts and public fountains throughout its empire to insure a clean supply of water for its citizens. Rome had a "water commissioner" who was responsible for seeing that the water supply was kept adequate and clean; the punishment for pollution of the water supply was death. Lead was commonly used for roman pipes and the subsequent fall of the roman empire has been related by some to the effects of lead on the roman brain.

Most ancient peoples recognized that some diseases were communicable and isolated individuals thought to carry "infections". An example of this is the universal shunning of lepers, which occurs even today. When the black death struck Europe, entire villages were abandoned as people fled in an effort to escape the highly infectious plague.

Similarly, in the middle ages the rich of Europe fled to their country homes when small pox struck in an effort to escape its terrible consequences. The fact that people who recovered from a particular disease were immune to that disease was probably recognized many different times in many places. Often these survivors were expected to nurse the ill. Greek and roman physicians routinely prescribed diet and exercise as a treatment for ills.

Sadly, we know that this knowledge did not help most of our ancestors and that the human life span was, until the last 200 years, more often than not cut short due to infectious disease. Even today approximately 15,000,000 children die per year, mainly from infectious diseases that are preventable with basic sanitation, immunization and simple medical treatments. One might honestly question just how far we have come in our treatment of disease. An excellent synopsis of the history of microbiology ancient people had certainly seen masses of microbes, such as mold and bacterial colonies, on spoiled food, but it is doubtful if anyone considered that they were viewing living organisms.

Small boys and maybe a few love-sick adults staring into a clear pond, must have seen tiny specks moving rapidly about and some may have considered them living creatures, but to express this to their friends would be equivalent to us telling our friends that we'd seen a flying saucer.

The first person to report seeing microbes under the microscope was an Englishman, robert hooke. Working with a crude compound microscope he saw the cellular structure of plants around 1665. He also saw fungi which he drew. However, because his lens were of poor quality he was apparently unable to "see" bacteria.

Anton van leeuwenhoek was a man born before his time. Although not the first to discover the microscope or to use magnifying lens, he was the first to see and describe bacteria. We know that he was a "cloth merchant" living in delft Holland. And that he used magnifying lens to view the quality of the weave of the merchandise he purchased. He travelled to England in 1668 to view English cloth and there he saw drawings of magnifications of cloth much greater than any of the current lens available in holland would do.

He returned to holland and took up lens grinding. Being meticulous, he developed his lens grinding to an art and in the process tested them by seeing how much detail he could observe with a given lens. One can guess that he chanced to look at a sample of pond water or other source rich in microbes and was amazed to see distinct, uniquely shaped organisms going, apparently purposefully, about their lives in a tiny microcosm. He made numerous microscopes from silver and gold and viewed everything he could including the scum on his teeth and his semen.

His best lens could magnify ~300-500 fold which allowed him to see microscopic algae and protozoa and larger bacteria. He clearly had excellent eyesight because he accurately drew pictures of microbes that were at the limit of the magnification of his lens. He used only single lens and not the compound lens of the true microscopes we employ today; which makes his observations all the more amazing. He wrote of his observations to the royal society of London in 1676 and included numerous drawings.

He astonished everyone by claiming that many of the tiny things he saw with his lens were alive because he saw them swimming purposefully about. This caused no end of shock and wonderment and numerous people hurried to delft to see if this dutchman was "in his cups" or if he was really onto something new and wonderful.

A few minutes with one of his numerous microscopes was all it took to convert his visitors to enthusiastic believers in the existence of these tiny beasties living all around them. His discovery was the equivalent of our finding life on mars today.

SPONTANEOUS GENERATION

The mystery of life has puzzled and confounded humans since the first human began to contemplate his world. The religions of ancient societies were built around the seasons, the sun and the renewal of life as these were so clearly tied to survival; both of the human species through birth and death and of the individual in the attainment of sustenance. Spontaneous generation or the idea that life routinely arises from non-life was a common sense explanation of the miracle of life.

It had the advantage of simplicity, ease of understanding and didn't require any waste-of-time thinking. As science and the scientific method grew with the slow accumulation of knowledge, observant individuals began to consider the origin of life more deeply. Simple observations convinced many people that all the larger animals and plants produced life from previous life. Despite this, the mass of humans clung to the comfortable idea of spontaneous generation further, religions saw it as a convenient way to demonstrate the hand of god operating continuously in the world.

Some individuals, such as j.b. Van helment even described how one could make mice from grain, a jar and dirty rags by putting them together in a dark place for a few weeks and soon mice would appear in the jar.

Other, more perceptive individuals, like f. Redi tested the common idea that maggots arise via spontaneous generation on rotting meat. He placed a piece of meat in three jars, one he left open, one he corked tightly and the third he covered with a fine mesh gauze. Maggots only appeared in the open container, no matter how long he left the jars. Redi's experiment was important because of its eloquent simplicity. Anyone could repeat it and obtain the same clear results. Nevertheless, many people clung fiercely to the idea of spontaneous generation, while others designed experiments to test it.

In every case the results of the majority of these experiments indicated that spontaneous generation did not occur. The intellectual ferment this controversy stirred up gradually evolved into the scientific method as the various antagonists questioned each other's assumptions and, more importantly, their experimental design.

These arguments forced the designing of better experiments and eventually persuaded all but the most recalcitrant believers to discard spontaneous generation as an explanation for all higher life forms. Then, in the 1800's the refinement of the microscope, through which people could see tiny life forms that they assumed were simple, gave spontaneous generation proponents new life.

Again, flawed common sense led reasonable people astray. As the existence of microscopic life was accepted, the assumption was that such life must be simple compared to higher, more complex life. The reasoning that followed this erroneous assumption was that since the microbes were small they must be simple and it followed that they were formed by spontaneous generation, hence god was still at work creating micro-life. Small is not simple!

The battle over spontaneous generation raged anew both from the pulpit and the lab. A number of scientists performed elementary experiments in which they treated soups and broth's, which left unheated would team with microbes after a few days, with heat to destroy any life present in them and asked the question: "Would new life arise in these sterile soups"? Spallanzani boiled "Soup" in glass containers and melted the glass closed.

The observation that nothing subsequently grew in this "heated" soup suggested that spontaneous generation didn't work. His detractors, rightly criticized his experiments, proposing that since air is necessary for life and since he had sealed the flask to air, obviously no life could develop. Others boiled soups and microbes grew, thus apparently supporting spontaneous generation. But again the preponderance of data suggested that spontaneous generation did not even apply to the "simple" microbes.

In 1859 one of the fathers of modern microbiology. Pasteur decided to settle the question of spontaneous generation once and for all. A genius at devising definitive experiments, pasteur first drew the necks of glass flasks out so that they remained open to the air, but were bent so that air could only enter by a curved path. He then added broth and boiled it to destroy contaminating microbes. These flasks were then incubated and observed for months. He reasoned that the microbes in the air that could contaminate the sterile broth would be trapped on the sides of the thin glass necks before they reached the sterile broth. If spontaneous generation didn't occur no growth should take place.

This is exactly what happened, the flasks remained sterile indefinitely, until pasteur tipped the sterile broth up into the curved neck where he predicted the airborne organisms would have settled. After doing this the broths always grew microbes. These experiments ended the spontaneous generation controversy because these experiment was so elegant and simple and the results so clear, that anyone could repeat them.

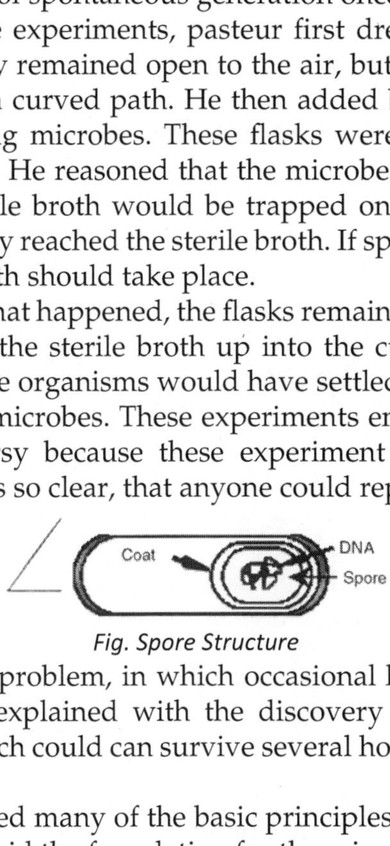

Fig. Spore Structure

Later, an earlier problem, in which occasional heated-broths did not remain sterile, was explained with the discovery of the heat resistant bacterial some of which could can survive several hours of boiling without being killed.

Pasteur discovered many of the basic principles of microbiology and, along with R. Koch, laid the foundation for the science of microbiology. In 1857 Napoleon III was having trouble with his sailors mutinying because their wine was spoiling after only a few weeks at sea. Naturally Napoleon was distraught because his hopes for world conquest were being scuttled (pardon the pun) over a little spoiled wine, so he begged Pasteur for help.

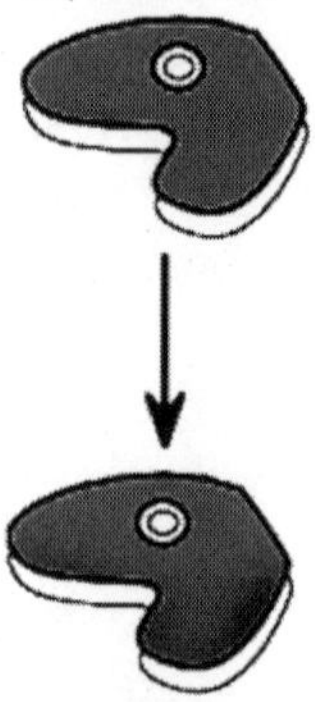

Fig. The Redi Experiment

Pasteur, armed with his trusty microscope, accepted the challenge and soon recognized that by looking at the spoiled wines he could distinguish between the contaminants that caused the spoilage and even predict the taste of the wine solely from his microscopic observations. He then reasoned that if one were to heat the wine to a point where its flavour was unaffected, but the harmful microbes were killed it wouldn't spoil.

As we are aware this process, today known as pasteurization, worked exactly the way he predicted and is the foundation of the modern treatment of bottled liquids to prevent their. It is important to realise that pasteurization is not the same as sterilization. Pasteurization only kills organisms that may spoil the product, but it allows many microbes to survive, whereas sterilization kills all the living organisms in the treated material. Pasteur also realized that the yeast that was present in all the produced the alcohol in wine.

When he announced this, a number of famous scientists were enraged, because the current theory of wine production was that wine formation was the result of spontaneous chemical changes that occurred in the grape juice. Pasteur was attacked furiously at scientific meetings, to the point where certain scientists did humorous skits about Pasteur and his tiny little yeast "stills" turning out alcohol. Pasteur had the last laugh however as people all over the world soon realized that if he was right they could control the quality of wine by controlling the yeast that made it. In a short period many others verified his observations and the opposition sank without a sound.

THE SPECTRUM OF MICROBIOLOGY

Like all other living things, microorganisms are placed into a system of classification. Classification highlights characteristics that are common among certain groups while providing order to the variety of living things. The science of classification is known as taxonomy, and taxon is an alternative expression for a classification category. Taxonomy displays the unity and diversity among living things, including microorganisms. Among the first taxonomists was Carolus Linnaeus. In the 1750s and 1760s, Linnaeus classified all known plants and animals of that period and set down the rules for nomenclature.

CLASSIFICATION SCHEMES

The fundamental rank of the classification as set down by Linnaeus is the species. For organisms such as animals and plants, a species is defined as a population of individuals that breed among themselves. For microorganisms, a species is defined as a group of organisms that are 70 per cent similar from a biochemical standpoint. In the classification scheme, various species are grouped together to form a genus. Among the bacteria, for example, the species Shigella boydii and Shigella flexneri are in the genus Shigella because the organisms are at least 70 per cent similar. Various genera are then grouped as a family because of similarities, and various families are placed together in an order. Continuing the classification scheme, a number of orders are grouped as a class, and several classes are categorized in a single phylum or division. The various phyla or divisions are placed in the broadest classification entry, the kingdom. Numerous criteria are used in establishing a species and in placing species together in broader classification categories. Morphology and structure are considered, as well as cellular features, biochemical properties, and genetic characteristics. In addition, the antibodies that an organism elicits in the human body are a defining property. The nutritional format is considered, as are staining characteristics.

PROKARYOTES AND EUKARYOTES

Because of their characteristics, microorganisms join all other living organisms in two major groups of organisms: prokaryotes and eukaryotes. Bacteria are prokaryotes because of their cellular properties, while other microorganisms such as fungi, protozoa, and unicellular algae are eukaryotes. Viruses are neither prokaryotes nor eukaryotes because of their simplicity and unique characteristics.

THE FIVE KINGDOMS

The generally accepted classification of living things was devised by Robert Whittaker of Cornell University in 1969. Whittaker suggested a five-kingdom classification. The first of the five kingdoms is Monera. Prokaryotes, such as bacteria and cyanobacteria are in this kingdom; the second kingdom, Protista, includes protozoa, unicellular algae, and slime molds, all of which are eukaryotes and single-celled; in the third kingdom, Fungi, are the molds, mushrooms, and yeasts. These organisms are eukaryotes that absorb simple nutrients from the soil. The remaining two kingdoms are Plantae and Animalia.

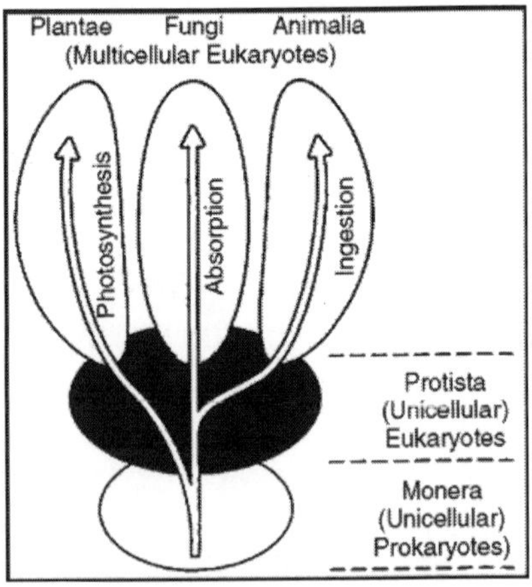

Fig. The Five Kingdoms of Living things Showing the Position of Microorganisms Relative to Other Organisms.

BRIEF DESCRIPTIONS OF MICROORGANISMS

Bacteria are relatively simple, prokaryotic organisms whose cells lack a nucleus or nuclear membrane. The bacteria may appear as rods, spheres, or spirals. Bacteria reproduce by binary fission, have unique constituents in their cell walls, and exist in most environments on earth. For instance, they live at temperatures ranging from 0° to 100°C and in conditions that are oxygen rich or oxygen free. A microscope is necessary to see and study them. Fungi are eukaryotic microorganisms that include multicellular molds and unicellular yeasts. The yeasts are slightly larger than bacteria and are used in alcoholic fermentations and bread making. Certain yeasts such as Candida albicans are pathogenic. Molds are filamentous, branched fungi that use spores for reproduction. The fungi prefer acidic environments, and most live at room temperature under oxygen-rich conditions. The common mushroom is a fungus.

Protozoa are eukaryotic, unicellular organisms. Motion is a characteristic associated with many species, and the protozoa can be classified just as to how they move: Some protozoa use flagella, others use cilia, and others use pseudopodia. Certain species are non-motile. Protozoa exist in an infinite variety of shapes because they have no cell walls. Many species cause such human diseases as malaria, sleeping sickness, dysentery, and toxoplasmosis. The term algae implies a variety of plantlike organisms. In microbiology, several types of single-celled algae are important. Examples are the diatoms and dinoflagellates that inhabit the oceans and are found at the bases of marine food chains. Most algae capture sunlight and transform it to the chemical energy of carbohydrates in the process of photosynthesis. Viruses are ultramicroscopic bits of genetic material enclosed in a protein shell and, sometimes, a membranous envelope. Viruses have no metabolism; therefore, it is difficult to use drugs to interfere with their structures or activities. Viruses multiply in living cells and use the chemical machinery of the cells for their own purpose. Often, they destroy the cells in the process of replicating.

NOMENCLATURE OF MICROORGANISMS

The system for naming all living things, established by Linnaeus, is also applied to microorganisms. In this system, all organisms are placed into a classification system, and each organism is given a binomial name. The binomial name consists of two names. The first name is the genus to which the organism belongs. The second name is a modifying adjective called the species modifier. In writing the binomial name, the first letter of the genus name is capitalized, and the remainder of the genus name and the complete species modifier are written in lowercase letters. The entire binomial name is either italicized or underlined. It can be abbreviated by using the first letter of the genus name and the full species modifier. An example of a microbial name is Escherichia coli, the bacterial rod found in the human intestine. The name is abbreviated E. coli.

CHEMICAL PRINCIPLES OF MICROBIOLOGY

In the 1700s, scientists discovered the chemical and physical basis of living things, and soon they realised that the chemical organization of all living things is remarkably similar. Microorganisms, as forms of living things, conform to this principle and have a chemical basis that underlies their metabolism.

ELEMENTS AND ATOMS

All living things on earth, including microorganisms, are composed of fundamental building blocks of matter called elements. Over 100 elements are known to exist, including certain ones synthesized by scientists. An element is a substance that cannot be decomposed by chemical means. Such things as oxygen, iron, calcium, sodium, hydrogen, carbon, and nitrogen are elements.

Each element is composed of one particular kind of atom. An atom is the smallest part of an element that can enter into combinations with atoms of other elements. Atoms consist of positively charged particles called protons surrounded by negatively charged particles called electrons. A third particle called the neutron has no electrical charge; it has the same weight as a proton. Protons and neutrons adhere tightly to form the dense, positively charged nucleus of the atom. Electrons spin around the nucleus in orbits, or shells.

The arrangement of electrons in an atom plays an essential role in the chemistry of the atom. Atoms are most stable when their outer shell of electrons has a full quota, which may be two electrons or eight electrons. Atoms tend to gain or lose electrons until their outer shells have this stable arrangement. The gaining or losing of electrons contributes to the chemical reactions in which an atom participates.

MOLECULES

Most of the microbial compounds of interest to biologists are composed of units called molecules. A molecule is a precise arrangement of atoms from different elements; a compound is a mass of molecules. The arrangements of the atoms in a molecule account for the properties of a compound. The molecular weight is equal to the atomic weights of the atoms in the molecule. For example, the molecular weight of water is 18. The atoms in molecules may be joined to one another by various linkages called bonds.

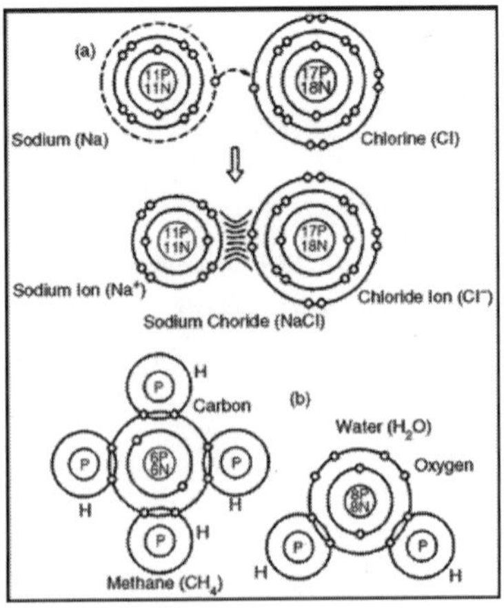

Fig. Bond Formation in Molecules.

(a) Formation of an ionic bond in a sodium chloride molecule.

(b) Covalent bonding in methane and water molecules.

One example of a bond is an ionic bond, which is formed when the electrons of one atom transfer to a second atom, creating electrically charged atoms called ions. The electrical charges attract the ions to one another; the attraction creates the ionic bond. Sodium chloride consists of sodium ions and chloride ions joined by ionic bonds. A second type of linkage is called a covalent bond which forms when two atoms share one or more electrons with one another. For example, carbon shares its electrons with four hydrogen atoms, and the resulting molecule is methane (CH_4). If one pair of electrons is shared, the bond is a single bond; if two pairs are shared, then it is a double bond. Covalent bonds are present in organic molecules such as proteins, lipids, and carbohydrates.

ACIDS AND BASES

Certain chemical compounds release hydrogen ions when the compounds are placed in water. These compounds are called acids. For example, when hydrogen chloride is placed in water, it releases its hydrogen ions, and the solution becomes hydrochloric acid. Certain chemical compounds attract hydrogen ions when they are placed in water. These substances are called bases. An example of a base is sodium hydroxide (NaOH). When this substance is placed in water, it attracts hydrogen ions, and a basic (or alkaline) solution results.

ORGANIC COMPOUNDS

The chemical compounds of living things such as microorganisms are known as organic compounds because of their association with organisms. The organic compounds, the subject matter of organic chemistry, are the compounds associated with life processes in microorganisms.

CARBOHYDRATES

Four major categories of organic compounds are found in all microorganisms. The first category is the carbohydrates. Carbohydrates are used by microorganisms as sources of energy. In addition, carbohydrates serve as structural materials such as in the construction of the microbial cell wall. Carbohydrates are molecules composed of carbon, hydrogen, and oxygen; the ratio of hydrogen atoms to oxygen atoms is 2:1. The simple carbohydrates are commonly referred to as sugars. Sugars are monosaccharides if they are composed of single molecules and disaccharides if they are composed of two molecules. The most important monosaccharide is glucose, a carbohydrate with the molecular formula $C_6H_{12}O_6$. Glucose is the basic form of fuel for many species of microorganisms.

It is soluble and is transported by body fluids to all cells, where it is metabolized to release its energy. Glucose is the starting material for cellular respiration, and it is the main product of photosynthesis in microorganisms. Three important disaccharides are also found in living things. One disaccharide is maltose, a combination of two glucose units covalently linked. Yeast cells break down the maltose from grain starch in the process of alcoholic fermentation. Another disaccharide is sucrose, the table sugar formed by linking glucose to another monosaccharide called fructose.

A third disaccharide is lactose, composed of glucose and galactose units. Lactose, the major carbohydrate in milk, is digested to acid by microorganisms when they sour milk and form sour-milk products such as yogurt and sour cream. Complex carbohydrates are known as polysaccharides. Polysaccharides are formed by linking eight or more monosaccharide molecules.

Among the most important polysaccharides are starches, which are composed of hundreds or thousands of glucose units linked to one another. Starches serve as a storage form for carbohydrates. Microorganisms break down starch to use the glucose it contains for their energy needs. Another important polysaccharide is glycogen, which is related to starch. Many bacteria have glycogen in their cytoplasm. Still another is cellulose. Cellulose is also composed of glucose units, but the units cannot be released from one another except by a few species of microorganisms, especially those in the stomach of the cow and other ruminants. The cell walls of algae contain cellulose, and certain fungi have this polysaccharide. Another polysaccharide called chitin is a primary constituent in the fungal cell wall.

LIPIDS

Lipids are organic molecules composed of carbon, hydrogen, and oxygen atoms. In contrast to carbohydrates, the ratio of hydrogen atoms to oxygen atoms is much higher. Lipids include steroids, waxes, and the most familiar lipids, fats.

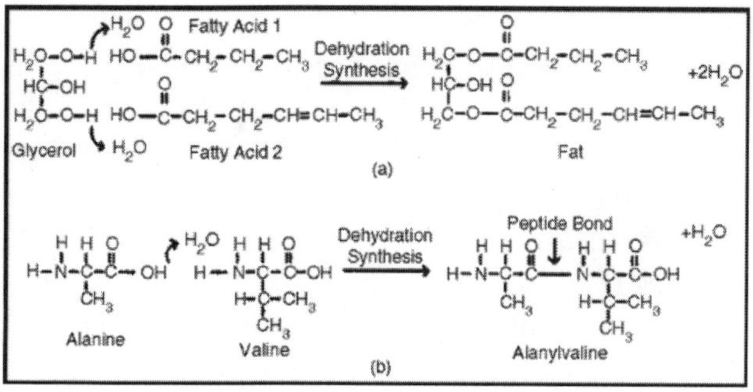

Fig. Syntheses in Organic Molecules.

(a) Bonding of two fatty acids to a glycerol molecule in the formation of a fat.

(b) Bonding of two amino acids via a peptide bond in the formation of a protein.

Fat molecules are composed of a glycerol molecule and one, two, or three molecules of fatty acids. A fatty acid is a long chain of carbon atoms with associated hydroxyl (–OH) groups. At one end of the fatty acid is an organic acid (–COOH) group. The fatty acids in a fat may be all alike or all different. They are bound to the glycerol molecule during dehydration synthesis, a process that involves the removal of water. The number of carbon atoms in a fatty acid may be as few as four or as many as 24.

Certain fatty acids have one or more double bonds in their molecules. Fats that include these molecules are called unsaturated fats. Other fatty acids have no double bonds. Fats that include these fatty acids are called saturated fats. Some microbial species use fats as energy sources. They produce the enzyme lipase, which breaks down fats to fatty acids and glycerol. An important type of phosphorus-containing lipid, the phospholipid, is a major constituent of the cell membranes of all microorganisms.

PROTEINS

Proteins are among the most complex of all organic compounds. They are composed of units called amino acids, which contain carbon, hydrogen, oxygen, and nitrogen atoms. Certain amino acids also have sulphur atoms, phosphorus, or other trace elements such as iron or copper. Many proteins are immense and complex as compared to carbohydrates or fats. However, all are composed of folded, long chains of the relatively simple amino acids. There are 20 kinds of amino acids, each with an amino ($-NH_2$) group and an organic acid (–COOH) group. The amino acids differ with respect to the nature of the chemical group that is attached to the base structure.

Examples of amino acids are alanine, valine, glutamic acid, tryptophan, tyrosine, and histidine. Amino acids are linked to form a protein by the removal of water molecules.

The links forged between the amino acids are called peptide bonds, and small proteins are often called peptides. All living things, including microorganisms, depend upon proteins for their existence. Proteins are the major molecules from which microorganisms are constructed. Certain proteins are dissolved or suspended in the watery substance of the cells, while others are incorporated into various structures of the cells, such as the cell membrane.

Bacterial toxins (metabolic poisons) and microbial flagella and pili are usually composed of proteins. An essential use for proteins is in the construction of enzymes. Enzymes catalyse the chemical reactions that take place within microorganisms. The enzymes are not used up in the reaction, but remain available to catalyse succeeding reactions. Without enzymes, the metabolic activity of the microorganism could not take place. Every species manufactures proteins unique to that species.

The information for synthesizing these unique proteins is found in the nucleus of the cell. The so-called genetic code specifies the sequence of amino acids in the protein and thereby regulates the chemical activity taking place within the cell. Proteins also can serve as a reserve source of energy for the microorganism. When the amino group is removed from an amino acid, the resulting compound is energy rich.

NUCLEIC ACIDS

Like proteins, nucleic acids are very large molecules. The nucleic acids are composed of smaller units called nucleotides. Each nucleotide contains a five-carbon carbohydrate molecule, a phosphate group, and a nitrogen- containing molecule that has basic properties and is called a nitrogenous base. Microorganisms contain two important kinds of nucleic acids. One type is called deoxyribonucleic acid, or DNA. The other is known as ribonucleic acid, or RNA. DNA is found primarily in the nucleus of eukaryotic microorganisms (which have nuclei) and suspended in the cytoplasm of prokaryotic microorganisms (which lack nuclei). DNA is also located in plasmids, the tiny loops of DNA found in bacterial cytoplasm. RNA is found in both the nucleus (if present) and the cytoplasm of the microorganism. DNA and RNA differ from one another in their components. DNA contains the carbohydrate deoxyribose, while RNA has ribose. In addition, DNA contains the bases adenine, cytosine, guanine, and thymine, while RNA has adenine, cytosine, guanine, and uracil.

DISEASES AND DISORDERS IN MICROBIOLOGY

CONTRIBUTING FACTORS

Infectious disease results from a competition for supremacy between the parasite and the host. If the parasite overcomes the host, there is a change in the general state of good health and disease develops. Several contributing factors are involved in the establishment of infectious disease. These factors determine whether the infecting organism will survive in the body.

Portals of Entry

In order for a pathogen to gain access to the host, the pathogen must pass through a portal of entry. One of the most common portals of entry is the mucous membranes, especially those of the respiratory, gastrointestinal, and urogenital tracts. Another important portal of entry is the skin. Penetration of the skin occurs during a wound or by a hair follicle. When microorganisms penetrate below the skin, the portal of entry is said to be the parenteral route.

Dose

The dose of an organism refers to the number of microorganisms required to establish an infection. For some diseases, such as typhoid fever, the dose is a few hundred bacteria. For other diseases, such as cholera, the dose may be several million bacteria. The dose may be expressed as the LD50, which refers to the dose of microorganisms that will kill 50 per cent of the hosts it enters.

Invasiveness

Invasiveness is a property that encourages disease because it refers to the ability of pathogens to penetrate into the tissues. Those organisms that cause intestinal ulcers, such as Entamoeba histolytica, penetrate the tissue effectively. Tissue invasion often begins with adherence, the ability of pathogens to attach to the tissue by using structures such as pili. The presence of a capsule or glycocalyx encourages adherence because they are composed of sticky materials.

Capsules

Microorganisms that possess capsules are able to resist host defences by interfering with phagocytosis. Normally, the body uses white blood cells to engulf and destroy pathogens. However, toxic substances in the capsule are able to destroy the white blood cells before the white blood cells perform phagocytosis. The organism of pneumonia Streptococcus pneumoniae is well known for the toxic materials in its capsule. Many other pathogens also possess capsules.

ENZYMES AND TOXINS

Enzymes

Many pathogens produce a series of enzymes to help overcome body defences and establish themselves in the host. One example is leukocidins, a group of enzymes that destroy white blood cells. This destruction lessens the body's ability to perform phagocytosis. Other bacterial enzymes are hemolysins. These enzymes destroy red blood cells. Streptococci, staphylococci, and certain Clostridium species produce hemolysins. Coagulases are bacterial enzymes that clot the blood. These enzymes convert fibrinogen into fibrin, which forms the threads of a blood clot. The clot helps staphylococci avoid the body's phagocytes and contributes to its pathogenicity. Other important enzymes are streptokinase and hyaluronidase. Streptokinase is a streptococcal enzyme that dissolves blood clots. This activity helps the organism escape the body's attempt to wall off an infection. Hyaluronidase destroys hyaluronic acid, a polysaccharide that "cements" cells together in a tissue. Hyaluronidase thus permits organisms to spread through tissues and establish themselves at sites distant from that of the intial infection. Another enzyme, called collagenase, breaks down collagen in the connective tissues of muscles. It thereby encourages the spread of infection.

Toxins

Many bacteria are able to produce poisonous substances called toxins. Toxins act on the body's cells, tissues, and organs and interfere with important body processes, thereby interrupting normal body functions. Those microorganisms that produce toxins are said to be toxigenic. The condition in which toxins are produced is called toxemia. Two important types of toxins are exotoxins and endotoxins. Exotoxins are proteins produced by bacteria during their growth and liberated into their surrounding environment. Exotoxins are produced chiefly by Gram-positive bacteria, and the genes for this production are carried primarily on the plasmids. Various types of exotoxins exist. Neurotoxins interfere with the nervous system, while enterotoxins interfere with activities of the gastrointestinal tract.

In response to toxins, the body produces special antibodies called antitoxins, which unite with and neutralize the toxins, providing defence against disease. It is possible to immunize against the effects of exotoxins by injecting toxoids into individuals. Toxoids are preparations of exotoxins chemically treated to destroy their toxigenicity but retain their ability to elicit antibody formation in the body. Toxoids are currently available to protect against diphtheria and tetanus (the DT injection). Endotoxins are portions of the cell wall of Gram-negative bacteria. They consist primarily of lipopolysaccharides and are released when bacteria break apart during the process of lysis. Since lysis occurs during antibiotic therapy, the effects of endotoxins can bring about a worsening of symptoms during the recovery period. This condition is called endotoxin shock. It is accompanied by fever, chills, aches, and cardiovascular collapse.

PATHOGENIC VIRUSES

Because viruses lack metabolic capabilities, they rely on other means for overcoming body defences and causing disease. Viruses avoid body defences by multiplying within host cells, where antibodies and other components of the immune system cannot reach them. The effect occurring in host cells during viral invasion is referred to as the cytopathic effect. The cytopathic effect can develop when the virus alters the metabolism of the cell and prevents it from producing essential cellular components. Alternatively, the virus may induce cells to cling together in a large mass called a syncytium. In some cases, the virus causes the cell's lysosomes to release enzymes which then destroy the cell.

NON-SPECIFIC BODY DEFENCE

NON-SPECIFIC MECHANISMS OF DEFENCE

The body possesses many mechanisms that impart non-specific defence. The objectives of these mechanisms are to prevent microorganisms from gaining a foothold in the body and to destroy them if they penetrate to the deeper tissues.

Mechanical Barriers

Mechanical barriers at the portal of entry represent the first line of defence for the body. These defences are normally part of the body's anatomy and physiology. The skin is a representative example. The outermost layers of skin consist of compacted, cemented cells impregnated with the insoluble protein keratin. The thick top layer is impervious to infection and water. In the unbroken state, it usually is not penetrated by pathogens.

The mucous membranes of the urinary, respiratory, and digestive tracts are another example. They are moist and permeable, but their fluids, such as tears, mucus, and saliva, rid the membrane of irritants. Nasal hairs trap particles in the respiratory tract, and the fluids exert a flushing action. Cilia on the cells sweep and trap particles in the respiratory tract, and coughing ejects irritants.

Chemical Defences

Among the non-specific chemical defences of the body are the secretions of lubricating glands. The tears and saliva contain the enzyme lysozyme, which breaks down the peptidoglycan of the cell wall of Gram-positive bacteria. The lactic acid of the vagina imparts defence, and the extremely caustic hydrochloric acid of the stomach is a barrier to the intestine. Semen contains the antimicrobial substance spermine that inhibits bacteria in the male urogenital tract.

Genetic Barriers

The hereditary characteristics of an individual are a deterrent to disease as well. For example, humans suffer HIV infection because their T-lymphocytes have the receptor sites for the human immunodeficiency virus. Dogs, cats, and other animals are immune to this disease because they do not possess the genes for producing the receptor sites. Conversely, humans do not suffer canine distemper because humans lack the appropriate receptor sites for the virus that causes the disease.

Inflammation

Inflammation is a non-specific response to any trauma occurring to tissues. It is accompanied by signs and symptoms that include heat, swelling, redness, and pain. Inflammation mobilizes components of the immune system, sets into motion repair mechanisms, and encourages phagocytes to come to the area and destroy any microorganisms present. Inflammation can be controlled by nervous stimulation and chemical substances called cytokines. These chemical products of tissue cells and blood cells are responsible for many of the actions of inflammation. The loss of fluid leads to a local swelling called edema. In some types of inflammation, phagocytes accumulate in the whitish mass of cells, bacteria, and debris called pus.

Fever

Fever is considered a non-specific defence mechanism because it develops in response to numerous traumas. Fever is initiated by circulating substances called pyrogens, which affect the brain's hypothalamus and cause the latter to raise the temperature. Although excessive fever can be dangerous, fever is believed to have a beneficial role because it retards the growth of temperature- sensitive microorganisms (for example, leprosy bacilli), and it increases the metabolism of body cells while stimulating the immune reaction and the process of phagocytosis.

Interferon

Interferon is a group of antiviral substances produced by body cells in response to the presence of viruses. Lymphocytes and macrophages produce alpha-interferon, epithelial cells produce beta-interferon, and T-lymphocytes produce gamma-interferon. The interferons do not directly inhibit viruses. Instead, they stimulate adjacent cells to produce substances that inhibit the replication of viruses in those cells. Interferons produced in response to one virus will protect against many other types of viruses, and for this reason, interferon is considered a non-specific form of defence.

PHAGOCYTOSIS

Phagocytosis is the cellular process of engulfing solid particles by the cell membrane to form an internal phagosome by phagocytes and protists. Phagocytosis is a specific form of endocytosis involving the vesicular internalization of solid particles, such as bacteria, and is, therefore, distinct from other forms of endocytosis such as the vesicular internalization of various liquids. Phagocytosis is involved in the acquisition of nutrients for some cells, and, in the immune system, it is a major mechanism used to remove pathogens and cell debris. Bacteria, dead tissue cells, and small mineral particles are all examples of objects that may be phagocytosed. The process is homologous to eating only at the level of single-celled organisms; in multicellular animals, the process has been adapted to eliminate debris and pathogens, as opposed to taking in fuel for cellular processes, except in the case of the Trichoplax.

In Immune System

Phagocytosis in mammalian immune cells is activated by attachment to Pathogen-associated molecular patterns (PAMPS), which leads to NF-ê Bactivation. Opsonins such as C_3b and antibodies can act as attachment sites and aid phagocytosis of pathogens. Engulfment of material is facilitated by the actin-myosin contractile system. The phagosome of ingested material is then fused with the lysosome, leading to degradation. Degradation can be oxygen-dependent or oxygen-independent.

- Oxygen-dependent degradation depends on NADPH and the production of reactive oxygen species. Hydrogen peroxide and myeloperoxidase activate a halogenating system, which leads to the destruction of bacteria.
- Oxygen-independent degradation depends on the release of granules, containing proteolytic enzymes such as defensins, lysozyme, and cationic proteins. Other antimicrobial peptides are present in these granules, including lactoferrin, which sequesters iron to provide unfavourable growth conditions for bacteria.

It is possible for cells other than dedicated phagocytes (such as dendritic cells) to engage in phagocytosis.

In Apoptosis

Following apoptosis, the dying cells need to be taken up into the surrounding tissues by macrophages in a process called Efferocytosis. One of the features of an apoptotic cell is the presentation of a variety of intracellular molecules on the cell surface, such as Calreticulin, Phosphatidylserine (From the inner layer of the plasma membrane), Annexin A1, and oxidised LDL. These molecules are recognised by receptors on the cell surface of the macrophage such as the Phosphatidylserine Receptor, or by soluble (free floating) receptors such as Thrombospondin 1, Gas-6, and MFG-E8, which themselves, then, bind to other receptors on the macrophage such as CD36 and Alpha-V Beta-3 Integrin.

In Protists

In many protists, phagocytosis is used as a means of feeding, providing part or all of their nourishment. This is called phagotrophic nutrition, as distinguished from osmotrophic nutrition, which takes place by absorption.

- In some, such as amoeba, phagocytosis takes place by surrounding the target object with pseudopods, as in animal phagocytes. In humans, Entamoeba histolytica can phagocytose red blood cells. This process is known as "erythrophagocystosis", and is considered the only reliable way to distinguish Entamoeba histolytica from non- invasive species such as Entamoeba dispar.
- Ciliates also engage in phagocytosis. In ciliates there is a specialized groove or chamber in the cell where phagocytosis takes place, called the cytostome or mouth.

The resulting phagosome may be merged with lysosomes containing digestive enzymes, forming a phagolysosome. The food particles will then be digested, and the released nutrients are diffused or transported into the cytosol for use in other metabolic processes. Mixotrophy can involve phagotrophic nutrition and phototrophic nutrition.

Several Stages of Phagocytosis

- Chemotaxis: This is the movement of cells up a gradient of chemotactic factors. It may be directly induced by a substance such as C5a, produced as a result of complement activation. It can also be indirectly induced as a consequence of release of preformed mediators within mast cells by the action of C3a or C5a e.g. eosinophil chemotactic factor, or neutrophil chemotactic factor. Leukotrienes, produced by the metabolism of mast cell arachidonic acid, are also chemotactic.

- Adherence: This works reasonably well for whole bacteria or viruses, but less so for proteins or encapsulated bacteria. In order to deal more effectively with encapsulateed bacteria, antibodies directed against the capsule enable the phagocytic cells to ingest the organisms, using their Fc receptors.
- Pseudopodium Formation: This is the protrusion of membranes to flow round the "prey".
- Phagosome Formation: Fusion of the pseudopodium with a membrane enclosing the "prey" leads to the formation of a structure termed a phagosome.
- Phago-lysosome Formation: The phagosome moves deeper into the cell, and fuses with a lysosome, forming a phago-lysosome. These contain hydrogen peroxide, active oxygen species (free radicals), peroxidase, lysozyme and hydrolytic enzymes. This is known as the oxidative burst, and leads to digestion of the phagolysosomal contents, after which they are eliminated by exocytosis. Some peptides however, undergo a very important separate process at this stage. Instead of being eliminated, they attach to a host molecule called MHC class II and end up being expressed on the surface of the cell within a groove on the MHC molecule (antigen presentation). The speed of phagocytosis can be increased markedly by bringing into action two attachment devices present on the surface of phagocytic cells:
  - Fc receptor: Which binds the Fc portion of antibody molecules, chiefly IgG. The IgG will have attached the organism via its Fab site.
  - Complement receptor: The third component of complement (C3) also binds to organisms and then attaches to the complement receptor.

THE IMMUNE SYSTEM

Immunity is a state of specific resistance to infection. Specific resistance is directed against a particular type of microorganism and is the single most important characteristic of immunity. The immune system enables the body to recognize a foreign agent as non-self, which is something other than a person's own substances (self). The immune system takes a specific action for neutralizing, killing, and eliminating that agent. The action involves non-specific resistance as well. On occasion, the immune system activity may lead to tissue damage as seen in allergic disorders and other states of hypersensitivity. The immune system's activity is based on its ability to distinguish characteristic proteins or protein-linked components associated with alien substances. Once this distinction has been made, certain lymphocytes are provoked to produce antibodies directed against the foreign matter, while other lymphocytes are sensitized to the invading agent and react with it directly. Thus, there are two major branches of the immune system: antibody-mediated immunity (also known as humoural immunity) and cell-mediated immunity.

ANTIGENS

Immune responses are directed at a series of foreign substances known as antigens, also referred to as immunogens. Most antigens are high molecular weight substances, but low molecular weight substances will also act as antigens if they bind to proteins in the body. The low molecular weight compound is referred to as a hapten. The resulting conjugate may induce an immune response directed against the antigen. The uptake and processing of antigens by macrophages in the tissue is an initial, critical step in most immune responses. The simple act of taking foreign substances into the body does not necessarily invoke an immune response because the substances may be broken down before they are ingested by macrophages.

Antigenic Determinants

The chemical groups on the antigen molecules that determine their immunogenicity are called antigenic determinants, also known as epitopes. Antigenic determinants may consist of several amino acids of a protein molecule or several monosaccharide units of a polysaccharide. Each species of living thing is chemically and antigenically unique because of differences in its proteins, carbohydrates, and other organic substances.

Types of Antigens

Certain types of antigens are distinctive. Autoantigens, for example, are a person's own self antigens. Alloantigens are antigens found in different members of the same species (the red blood cell antigens A and B are examples). Heterophile antigens are identical antigens found in the cells of different species. A single organism such as a bacterium may contain a variety of proteins, carbohydrates, and other materials that provoke immune responses. Antigens found on the body cell are called somatic antigens. Antigens in the bacterial capsule are capsular antigens. Antigens of an organism's flagella are known as flagellar antigens (H antigens). Protein substances such as exotoxins are also antigenic.

CELLS OF THE IMMUNE SYSTEM

Cells of the immune system are associated with the lymphatic system of the body and its specialized cells. Lymphocytes of the lymphatic system are derived from stem cells of the bone marrow. These undifferentiated precursor cells proliferate throughout life and replenish the mature cells of the immune system.

B-lymphocytes and T-lymphocytes

There are two major pathways for the differentiation of stem cells into immune cells. Certain of the stem cells produce B-lymphocytes (B-cells) while other stem cells form T-lymphocytes (T-cells). B-lymphocytes are so named because in birds, they are formed in the bursa of Fabricius. The equivalent site in humans has not been identified but is believed to be the bone marrow. T- lymphocytes undergo their conversion in the thymus gland, an organ in the neck tissues near the trachea and thyroid gland.

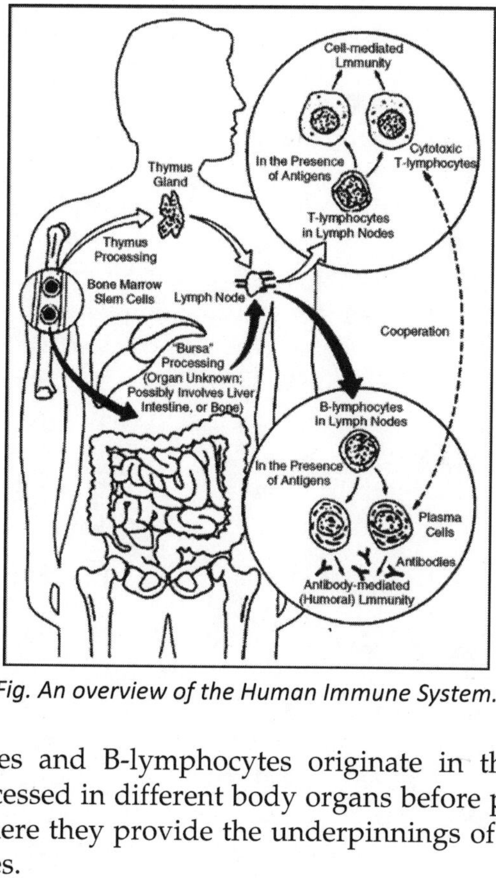

Fig. An overview of the Human Immune System.

T-lymphocytes and B-lymphocytes originate in the bone marrow and then are processed in different body organs before proceeding to the lymph nodes, where they provide the underpinnings of the two types of immune responses.

The transformation of stem cells into B-lymphocytes and T-lymphocytes begins about the fifth month after fertilization, and a full set is complete a few months after birth. These cells then migrate to the lymphoid organs in the lymph nodes, spleen, tonsils, adenoids, and other organs of the lymphatic system. To initiate the immune response1 microorganisms are phagocytized and their antigens are processed in phagocytic cells such as macrophages. The antigenic determinants are displayed on the surface of the phagocytic cells and presented to the appropriate B-lymphocytes and T-lymphocytes to provoke an imn1une response.

Clonal Selection

The donal selection theory helps explain how lymphocytes recognize antigenic determinants and respond. Small populations (clones) of lymphocytes bear receptors on their cell membranes. Production of these receptors is genetically determined. On B-lymphocytes, the receptors consist of antibody molecules, whilP on T-lymphocytes, they are clusters of amino acids.

When lymphocytes encounter an antigenic determinant on the surface of a macrophage, their receptors n1atch vvith the antigenic detenninant and a stimulation follows. A match is also made between a set of molecules called major histocompatibility (MHC) molecules and their receptors. The clonal selection theory suggests that B-lymphocytes and T-lymphocytes exist for all antigenic determinants even before contact with an antigen is made.

The theory also says that antigenic determinants stimulate the lymphocytes to endow their progeny with identical specificity. The B lymphocytes and T-lymphocytes that might potentially react with the body's own cells are deleted or in some way inactivated to ensure that an immune response to the host organism does not develop. Two general types of immunity exist for specific resistance to disease. They are antibody-mediated (humoural) immunity, centred in B-lymphocytes, and cell-mediated immunity, centred in T-lymphocytes.

ANTIBODY-MEDIATED (HUMOURAL) IMMUNITY (AMI)

On exposure to antigenic determinants in lymphatic organs, B lymphocytes are activated and differentiated to form plasma cells. Plasma cells are specialized, differentiated cells that synthesize and secrete antibodies specific for an antigen. Other activated B-lymphocytes form memory cells. These cells can be activated later to differentiate to plasma cells for rapid antibody production.

This antibody production will occur on future re-entry of the antigen to the body and is the basis of long-term immunity. The products of plasma cells are antibodies. An antibody is a specialized protein substance produced by the host cells in response to an antigen in the host's tissues.

Antibodies are capable of reacting specifically with the antigen that provoked their production. Antibodies are often referred to as immunoglobulins. They circulate in the blood and are associated with the gamma globulin fraction of the plasma.

Structure of Antibodies

Structurally, the antibodies (immunoglobulins) are composed of four protein chains arranged in a distinctive pattern. Each molecule has two long chains of heavy molecular weight protein (H chains) and two short chains of light molecular weight protein (L chains). The chains are linked by sulphur bonds. At the outer end of each arm of the antibody molecule, a specific amino acid sequence exists. This is where the antibody molecule reacts with the antigenic determinant that provoked its production.

The combining site is known as the Fab region. The most common antibody molecules have two Fab regions and are said to be bivalent (having two combining sites). The remaining portion of the antibody molecule is called the Fc region because it can be crystallized. Its amino acid content is relatively constant and characteristic for its class. This portion of the molecule activates the complement system and encourages phagocytosis.

When the antibody molecule reacts with the antigen, the two surfaces fit together like the pieces of a jigsaw puzzle. This "recognition" is exquisitely accurate and accounts for the extreme specificity of antibody molecules.

Classes of Antibodies

Five classes of antibodies (immunoglobulins) are produced by the plasma cells. The first class, called IgM, is the major component of the primary antibody response in adult humans and is the first antibody to appear in the immune reaction. It is composed of five units joined by a J (joining) chain. IgM cannot diffuse through cell membranes and is found almost exclusively in the blood. Because of its many binding sites, it is more reactive with antigens than are other kinds of antibodies. IgM is also formed by the fetus during uterine development and is active against the A and B factors of the red blood cells. Many antitoxins formed against bacterial toxins are composed of JgM.

The principal antibody of the secondary immune response is IgG. This antibody is the most common in the bloodstream and is found in many secretions, such as spinal, synovial, lymph, and peritoneal fluids. IgG crosses the placenta and protects the fetus and newborn. IgG therefore provides a natural type of passive immunity. IgG also forms in the primary antibody reaction after a large amount of IgM has already formed. IgG has the "typical" antibody structure of four protein chains.

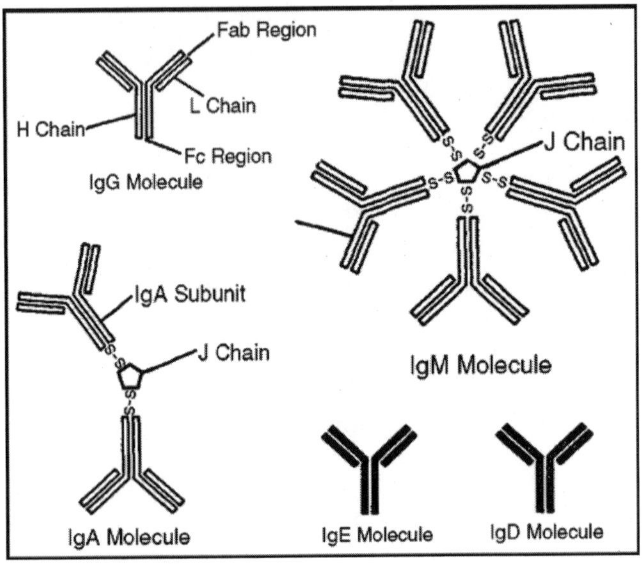

Fig. An Antibody Molecule and the Structures of the Five

Types of Antibody Molecules Produced by the Human Body.

The third class of antibody is IgA. This antibody is found in external secretions such as those at the mucosal surfaces of the respiratory, gastrointestinal, and urogenital tracts. It is also present in the tears, saliva, bile, urine, and colostrum, and it is transferred in the breast milk. IgA consists of two antibody units held together by a J chain and secretory component. IgA resists infections at the body surface.

The fourth class of antibody, IgD, is found in extremely small quantities in the serum. IgD is found at the surface of B-lymphocytes and is believed to be an antigen receptor at this location. IgM is also believed to be a receptor. The final kind of antibody, IgE, occurs in minute concentrations in the serum and is important in hypersensitivity reactions, especially the anaphylactic reaction.

A localized reaction is called allergy. The reaction of antibodies with antigens helps neutralize the antigen and restrict the spread of infection. Certain antibodies react with the bacterial surface, while others react with the flagella, pili, or capsules.

These reactions encourage phagocytosis. When antibodies react with a virus, viruses cannot attach to host cells and replicate. Antibody reaction with antigens also sets off the complement system, which results in the formation of an attack complex along with increased phagocytosis. Antibody reactions with toxin molecules neutralize the toxins and prevent further damage to body tissues.

Types of Immunity

Immunity may be broadly classified as innate or acquired. Innate immunity is present from birth. It consists of numerous types of non-specific factors that operate during times of disease. Acquired immunity is derived from activity of the immune system. The term generally refers to antibodies and is subdivided into two parts: active immunity and passive immunity. Active immunity is acquired when the body produces antibodies.

The immunity is usually long lasting because the immune system has been stimulated into action. However, it takes several hours to develop. Active immunity can be natural or artificial. Naturally acquired active immunity develops when a person produces antibodies during a bout of illness or on exposure to a microorganism even though disease does not occur. The B- lymphocytes and plasma cells function, and this immunity occurs during the "natural" scheme of events. Active immunity can also occur by artificial means. Artificially acquired active immunity occurs when a person produces antibodies after exposure to a vaccine. A vaccine consists of bacteria, viruses, or fragments of these. A vaccine may also contain toxoids, which are chemically treated bacterial toxins. Toxoid vaccines are available against diphtheria and tetanus. Viral vaccines are available against measles, mumps, rubella, polio, rabies, hepatitis A, hepatitis B, and yellow fever. Because vaccine exposures do not happen in the natural scheme of events, the immunity is said to be artificial. Passive immunity comes about when the body receives antibodies from an outside source. In passive immunity, the immune system does not operate and the immunity is not long lasting. However, it protects instantaneously. Passive immunity can be either natural or artificial.

Naturally aquired passive immunity develops when antibodies pass from mother to child across the placental lining. The newborn is born with IgG, and it receives maternal IgA if it is breast-fed. These antibodies will remain for approximately six months after birth before fading away. For artificially acquired passive immunity, a person is given an injection of antibodies. Usually these antibodies are derived from the blood of another individual who has produced antibodies when confronted with a certain disease.

The antibody injected is usually IgG. The antibodies will remain for a period of several days or weeks and then disappear. This type of immunity is given to protect people who have been exposed to tetanus, diphtheria, or botulism. It is also used to protect against certain serious viral diseases. Allergic reactions to the serum proteins (serum sickness) limit the use of the preparations.

CELL-MEDIATED IMMUNITY (CMI)

Cell-mediated immunity depends upon the activity of T-lymphocytes. T- lymphocytes have a longer life span than B-lymphocytes and are found in the same lymphatic tissues as the B-lymphocytes. The T-lymphocytes react with certain antigenic determinants and become immunologically "committed." Part of this commitment is the conversion to a subset of cells called cytotoxic T-lymphocytes.

Activity of Cytotoxic T-lymphocytes

Cytotoxic T-lymphocytes do not produce antibody molecules. Rather, they leave the lymphatic tissues and enter the circulation. They circulate through the blood vessels and gather at the infection site. Here they interact directly with organisms such as fungi, protozoa, cancer cells, and transplant cells. They also interact with virus-infected cells and bacteria-infected cells (such as lung cells infected with tuberculosis). The T-lymphocytes exert a "lethal hit" on the cells and secrete substances into them that lead to cellular destruction.

In addition to their direct interaction, T-lymphocytes also secrete substances called lymphokines. Lymphokines attract phagocytes to the area and encourage them to perform phagocytosis on fungi, protozoa, and infected cells. This activity helps relieve the infection. Lymphokines are also known as cytokines. An important cytokine is interleukin-1, which activates T- lymphocytes, causing them to proliferate further and form clones.

Helper and Suppressor T-lymphocytes

Helper T-lymphocytes also function in the immune system by encouraging the activity of B-lymphocytes in the production of antibodies. Suppressor T- lymphocytes regulate or suppress the activity of the immune system so that it is not excessive. Natural killer (NK) cells are T-lymphocytes that recognize and destroy many types of target cells without being exposed to antigens. Technically, these are not part of the specific immune response. Finally, the delayed hypersensitivity T-lymphocytes function in hypersensitivity reactions and encourage local tissue inflammations.

DETECTING ANTIBODIES WITH LABORATORY TESTS

Certain laboratory tests are available to detect the presence of antibodies in an individual. These laboratory tests are commonly used in diagnostic procedures because the presence of a certain antibody indicates the presence of a certain disease. The study of serum for its antibody content is known as serology. One of the objects of serology is to determine the titer of antibody present in the individual. The titer is an estimate of the antibody level in a unit volume of serum. It is determined as a reactive dilution of the serum and is usually expressed in a ratio such as 1:100.

The Agglutination and Precipitation Tests Serological tests that involve clumping of an antigen are called agglutination tests. Antibodies involved in agglutination tests are called agglutinins. When the antibody molecules unite with antigen molecules on the surface of bacteria, red blood cells, or particles, they cause the cells to stick together and form large clumps. Another form of serological test is the precipitation test. In this test, antibodies are called precipitins. They react with dissolved antigens and form large complexes that become visible as a fine precipitate. Tests such as these can be performed in fluid or gel.

The Complement Fixation Test

The complement fixation test is another application of a serological reaction. In this test, antibodies are combined with antigens in the presence of complement, a series of blood proteins. If the antibodies are specific for that antigen, they will react and use up ("fix") the complement. When a subsequent combination of blood cells and their antibodies are added to the tube, no complement will be available and no reaction will take place. However, if the first antibodies were not specific for their antigen, the complement is not fixed, and it is available to react with the blood cells and their antibodies. The blood cells will undergo lysis as a result of the complement and antibody activity.

The Neutralization Test

Serological tests that determine the presence of antitoxins are called neutralization reactions. Antitoxins are antibodies formed against toxins. When these antibodies are united with toxin molecules, they neutralize the toxins and will prevent their toxic effects in an animal. However, if the antitoxin is not specific for the toxin, the toxin is not neutralized and it will exert a lethal effect on the animal.

The Fluorescent Antibody Test

The fluorescent antibody technique is a serological test that helps to make visible an antigen-antibody reaction. Antibody solution is combined with cells that contain antigens. If an antibody-antigen reaction occurs, a fluorescent dye accumulates on the surface, thereby signaling a positive reaction.

Enzyme Immunoassays

In the enzyme immunoassay, an antigen-antibody reaction is encouraged on the surface of a particle, and an enzyme accumulates there. When a substrate is added, the enzyme molecules change the colour of the substrate, and the reaction can be designated positive. This test is sometimes called the enzyme- linked immunosorbentassay (ELISA). A contemporary development in immunology is the development of monoclonal antibodies.

Monoclonal antibodies are produced by a single colony of immortal cells, that is, cells with the ability to live and multiply for extremely long periods.

The cells producing the antibodies are hybrids formed by fusing antibody- producing plasma cells with tumor cells (the immortal cells). Monoclonal antibodies can be purified and used for diagnostic purposes as well as carriers for toxic chemical compounds. The process can possibly be used to kill tumor cells in the body if antitumor drugs are carried.

DISORDERS OF THE IMMUNE SYSTEM

HYPERSENSITIVITY REACTIONS

Certain human disorders are attributed to activity of the immune system. These disorders are commonly known as hypersensitivities, states of increased immune sensitivity that are mediated by antibody or cellular factors. The disorders may also involve immunodeficiencies in which failures of antibody- mediated or cell-mediated immunity take place. Normally the immune system plays an important role in protecting the body from microorganisms and other foreign substances.

If the activity of the immune system is excessive or overreactive, a hypersensitivity reaction develops. The consequences of a hypersensitivity reaction may be injury to the body or death. Most injury resulting from hypersensitivities develops after an interaction has taken place between antigens and antibodies or between antigens and sensitized T-lymphocytes.

The general nature of and symptoms accompanying the reaction depend upon whether antibodies or sensitized T-lymphocytes are involved. When antibodies are involved, the reactions fall under the heading of immediate hypersensitivity. When T-lymphocytes are involved, the reactions are characterized as delayed hypersensitivity. Immediate hypersensitivity reactions include anaphylaxis, allergic reactions, cytotoxic reactions, and immune complex reactions. Delayed hypersensitivity reactions are generally characterized as contact dermatitis or infection allergies.

Immediate Hypersensitivity

The reactions accompanying immediate hypersensitivity depend upon the nature of the antigen, the frequency and route of antigen contact, and the type of antibody reacting with the antigen. The initial dose of antigen is referred to as the sensitizing dose. This exposure is followed by a latent period and then a later dose of the same antigen, called the eliciting dose or shocking dose. The shocking dose sets off the hypersensitivity reaction, resulting in tissue damage. Immediate reactions begin within minutes of contact with the eliciting dose of antigen. If antigens are introduced directly into the tissues, such as by insect sting or injection, the result is a systemic reaction such as anaphylactic shock. When the contact is a superficial one involving the epithelial tissues, the reaction is more localized, as occurs in asthma or allergic rhinitis (hay fever). These local reactions are commonly referred to as allergy. Another term used is atopy. The antigens eliciting an immediate hypersensitivity are called allergens, particularly when they are involved in local allergic reactions.

Hapten molecules such as penicillin molecules may be involved when they are bound to larger protein molecules. Foods, feathers, pollen grains, animal dander, and dust may be allergens. Animal sera, bee venoms, and wasp venoms are also allergens. The antibodies involved in anaphylaxis reactions are of the type IgE. In cytotoxic and immune complex reactions, IgG and IgM are involved.

Anaphylaxis

Anaphylaxis, or type I hypersensitivity, is a whole-body, immediate hypersensitivity also known as anaphylactic shock. The allergens are introduced to the body directly to the tissues in a concentrated form (intramuscular or intravenous injection, for example). After the sensitizing dose has been administered, IgE is produced by the plasma cells. The antibodies circulate in the blood and attach at the Fc end to mast cells of the tissues and basophils in the bloodstream. This activity occurs during the latent period. When the eliciting dose of allergen is later administered, the antigens combine with antibodies on the surface of the mast cells and basophils.

After the antigen-antibody combination has taken place, the cells release a number of physiologically active substances including histamine and serotonin. These substances are derived from granules within the cell. The histamine, serotonin, and other mediators induce spasms of the smooth muscle, such as in the bronchioles, small arteries, and gastrointestinal tract lining.

A sudden drop in blood pressure occurs, followed by circulatory collapse and shock. Bronchospasms and edema cause constriction of the respiratory passageways, and breathing is very difficult. Facial edema occurs, and the heart rate increases due to constriction of the arteries. Swellings called "hives" develop at the site of injection and other areas of the skin. In severe cases, anaphylactic shock may result in death within several minutes to an hour. To relieve the symptoms, epinephrine is administered together with a smooth muscle relaxer, a drug such as cortisone to reduce swelling, and other drugs as appropriate.

Allergic Reactions

Allergic reactions (allergy) are a milder, localized form of anaphylaxis. As noted, such things as foods, pollen grains, and animal dander can induce these localized reactions. IgE, basophils, and mast cells are involved, but much less than in anaphylaxis. There appears to be a genetic basis for allergic reactions, as evidenced by their distribution in families.

Cytotoxic Reactions

Cytotoxic reactions are a form of immediate hypersensitivity, sometimes referred to as type II hypersensitivity. In these reactions, IgE and IgM are produced in response to stimulation by antigens. The antibodies unite with the antigens in the bloodstream, but they also unite with analogous antigens on the surface of the human body's cells. This union sets off the complement system, and destruction of the local tissue cells ensues.

An example of a cytotoxic reaction is thrombocytopenia. In this disease, antibody molecules are elicited by certain drug molecules. The antibodies unite with antigens on the surface of thrombocytes (platelets), and with complement activation, the thrombocytes are destroyed. The result is an impaired blood- clotting mechanism.Another example of the cytotoxic reaction is agranulocytosis. In this immune disorder, antibodies unite with antigens on the surface of neutrophils. As these cells are destroyed with complement activation, the capacity for phagocytosis is reduced. Cytotoxic reactions are also manifested by the transfusion reaction occurring when improper blood transfusions are performed. Another consequence is erythroblastosis fetalis, also known as hemolytic disease of the newborn, or Rh disease. In this condition, a pregnant woman produces Rh antibodies against the developing fetus, and when the Rh antibodies unite with Rh antigens on the surface of fetal red blood cells in a succeeding pregnancy, the red blood cells are destroyed.

Immune Complex Disease

Immune complexes are combinations of antigen and antibody that have the ability to fix complement. The antibodies involved are IgM or IgG, and the antigens exist in fluid as soluble antigens. Proteins or nucleic acids may be involved. An example of immune complex hypersensitivity is serum sickness. In this condition, animal serum is administered to humans, and its proteins elicit antibody production. When the antibodies and antigens unite, they form immune complexes, which activate the complement system and cause local tissue damage. The patient may display edema of the hands, face, and feet, as well as swelling of the upper respiratory tissues and impairment of normal respiration. An inflammatory response results. Formation of immune complexes is also involved with numerous diseases including systemic lupus erythematosus, rheumatoid arthritis, and glomerulonephritis. Immune complex hypresensitivity is often called type III hypersensitivity.

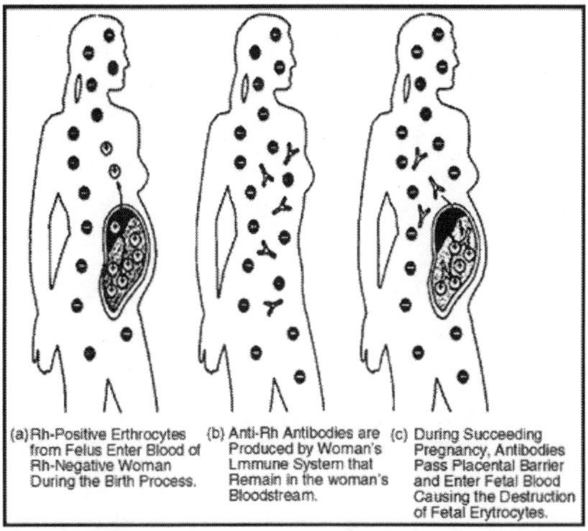

Fig. The Cytotoxic Reaction in Erythroblastosis Fetalis.

Delayed Hypersensitivity

T-lymphocytes rather than antibodies function in cases of delayed hypersensitivity, also called type IV hypersensitivity. Normally these are the T-lymphocytes involved in cell-mediated immunity. The T-lymphocytes produce lymphokines, which stimulate an influx of macrophages to perform phagocytosis. In delayed hypersensitivity, the result is an exaggeration of the immune response, and the phagocytes bring about the destruction of the local tissue.

Delayed hypersensitivity (also called cellular hypersensitivity) is so named because the reaction requires a day or more to develop. One manifestation of the reaction is infection allergy, as in the tuberculin skin test. A purified protein derivative (PPD) of Mycobacterium tuberculosis is applied to the skin superficially, and a skin reaction (swelling and redness) occurs 24 to 48 hours later if the person has had a previous exposure to the antigens of Mycobacterium tuberculosis, possibly during an episode of tuberculosis.

A second manifestation of delayed hypersensitivity is contact dermatitis. In many cases, the reaction is accompanied by large, blisterlike lesions in which vesicles are surrounded by a zone of erythema (redness). Usually, the vesicles itch intensely. Antigens involved in contact dermatitis include metals such as nickel and mercury, cosmetics, disinfectants, and plant substances such as the resins of poison ivy, poison oak, and poison sumac.

The individual can be tested to determine which antigen is the cause of allergy by performing a patch test. In this procedure, a patch containing a specific antigen is attached to the skin and left in place for 48 hours to determine if a reaction will take place.

IMMUNODEFICIENCY DISEASES

Disorders of the immune system may also be due to deficiencies of the system. These deficiencies may be congenital or iatrogenic (induced by immunosuppressive drugs), or they may result from malignancies occurring in the lymphatic system. Abnormalities of the B-lymphocytes result in an immunodeficiency accompanied by abnormal production of antibodies. These abnormalities may result from infections such as diseases of the liver, from immune complex disorders such as systemic lupus erythematosus, or from malignancies. Multiple myeloma is a malignant disease of plasma cells in which certain clones of B-lymphocytes produce an overabundance of antibodies. The antibodies are excreted into the urine, where they are known as Bence Jones proteins. Decreased ability of the B-lymphocytes to produce antibodies is called hypogammaglobulinemia.

Such a condition may result from a genetic defect or the failure of the body to produce B-lymphocytes. Bone marrow disorders may also be a cause. Abnormalities of the T-lymphocytes result in reduced capacity for cell- mediated immunity. This condition, called DiGeorge's syndrome, often develops from malformation or absence of the thymus gland.

An individual with this syndrome is highly susceptible to infection by fungi, protozoa, and viruses. When both B-lymphocytes and T-lymphocytes are deficient, the condition is called severe combined immunodeficiency. Recent procedures employing transplantation of bone marrow tissue have attempted to relieve this condition.

❐

Chapter - 2

Identification of Microorganism

A microorganism or microbe is an organism that is unicellular or lives in a colony of cellular organisms. The study of microorganisms is called microbiology, a subject that began with Anton van Leeuwenhoek's discovery of microorganisms in 1675, using a microscope of his own design. Microorganisms are very diverse; they include bacteria, fungi, archaea, and protists; microscopic plants; and animals such as plankton and the planarian. Some microbiologists also include viruses, but others consider these as non- living. Most microorganisms are unicellular but this is not universal, since some multicellular organisms are microscopic, while some unicellular protists and bacteria, like Thiomargarita namibiensis, are macroscopic and visible to the naked eye. Microorganisms live in all parts of the biosphere where there is liquid water, including soil, hot springs, on the ocean floor, high in the atmosphere and deep inside rocks within the Earth's crust.

Microorganisms are critical to nutrient recycling in ecosystems as they act as decomposers. As some microorganisms can fix nitrogen, they are a vital part of the nitrogen cycle, and recent studies indicate that airborne microbes may play a role in precipitation and weather. Microbes are also exploited by people in biotechnology, both in traditional food and beverage preparation, and in modern technologies based on genetic engineering. However, pathogenic microbes are harmful, since they invade and grow within other organisms, causing diseases that kill people, other animals and plants.

WHAT IS A MICROORGANISM?

The microorganism term is used instead of microbes; they are very small organisms size belonging to various groups: Bacteria, Fungi, Protozoa, one can add Viruses, characterized by their non-cellular structure. Microorganisms were observed for the first time in 1674 by Van Leeuwenhoek but they are only identified 200 years later.

Although those organisms are incredibly small (their size ranges from 0.03;un for the virus of the foot-and-mouth disease to 300 ;tm for Paramecium), they play a significant role to keep the planet running. Indeed, they carry out 90 per cent of the biochemical reactions produced on earth, without them the life on earth would be destroyed.

The micro-organisn1s are probably the first alive beings appeared on ground, they would be thus the common ancestors of all the living being. Although microorganisms appeared 3.8 billion years ago, they are observed by Pasteur and koch through their microscope way after their apparition. Pasteur delivered the fatal blow to the doctrine of spontaneous generation thanks to his examination and Koch discovered revealed the existence of bacilla at the origin of tuberculosis. Those discoveries changed radically the perception of the human beings on the world.

The chronology will give to the reader a global vision of all the discoveries during those past2 centuries. A microorganism (also spelled as microrganism) or microbe is an organism that is microscopic (too s1nall to be seen by the human eye). Microorganisms include bacteria, fungi, archaea or protists, but not "iiruses and prions, which are generally classified as non-living. Most microorganisms are single-celled, or unicellular, but son1e are microscopic, and some unicellular protists are visible to the average human. Microorganisms live ahnost everyvvhere on Earth where there is liquid vvater1 including hot springs, on the ocean floor, and deep inside rocks within Earth1s crust. Microorganisms are critical to nutrient recycling in ecosystems as they act as decomposers. As some microorganisms can also fix nitrogen, they are an important part of the nitrogen cycle. Hovvever, pathogenic microbes can invade other organisms and cause diseases that kill millions of people every year. Microorganisms can be found ahnost anyvvhere in the taxonomic organization of life on the planet. Bacteria and archaea are almost always microscopic, while a number of eukaryotes are also microscopic1 including n1ost protists and a number of fungi.

Viruses are generally regarded as not living and therefore are not microbes, although the fidel of microbiology also encompasses the study of viruses. Habitats and ecology Microorganisms are found in almost every habitat present in nature. Even in hostile environments such as the poles, deserts, geysers, rocks, and the deep sea, some types of microorganisms have adapted to the extrelne conditions and sustained colonies; these organisms are knov.rn as extremophiles. Extremophiles ha,./e been isolated from rocks as much as 7 kilometres beluw the earth's surface, and it has been suggested that the amount of living organisms belovv the eartl{s surface may be comparable with the amount of life on or above the surface. Extremophiks have been known to survive for a prolonged time in a vacuunl, and can be highly resistant to radiation, vvhich may even allovv thenl to survive in space. Many types of microorganisms have intimate symbiotic relationships with other larger organisms; some of which are mutually beneficial (mutualism), while others can be damaging to the host organism (parasitism). If microorganisms can cause disease in a host they are known as pathogens. Microorganisms are vital to humans and the environment, as they participate in the Earth's element cycles such as the carbon cycle and nitrogen cycle, as well as fulfilling other vital roles in virtually all ecosystems, such as recycling other organisms' dead remains and waste products through decomposition. Microbes also have an important place in most higher-order multicellular organisms as symbionts. Microorganisms are the cause of many infectious diseases.

The organisms involved include bacteria, causing diseases such as plague, tuberculosis and anthrax; protozoa, causing diseases such as malaria, sleeping sickness and toxoplasmosis; and also fungi causing diseases such as ringworm, candidiasis or histoplasmosis. However, other diseases such as influenza, yellow fever or AIDS are caused by viruses, which are not living organisms and are not therefore microorganisms. As of 2007, no clear examples of archaean pathogens are known, although a relationship has been proposed between the presence of some methanogens and human periodontal disease. Microorganisms are very tiny one-celled organisms, viruses, fungi, and bacteria, and are found everywhere in the world. They are found in all living things, plants and animal. There are more microorganisms on and inside your body than there are cells that make up your entire body. Microorganisms can live in the air, on land, and in fresh or salt water environments. Some of them, pathogens, can be harmful and causes diseases, but there are some microorganisms that are needed for living things to survive.

LAND MICROBES

All of the living things, plant and animal, in earth's environmental communities of forests, deserts, tundra, water, air, and all of the rest depend on the cryptobiotic crust or microbiotic layer in the soil. This is the layer of soil that most microbes live in. These microbe communities are made up of fungi, cyanobacteria and lichens. They look like a grayish cover on the ground when they are first forming, but do form in clumps of lichen that look like little hills after about 50 years of growth.

The cyanobacteria Nostoc lives on the land and forms in filaments of hyphae that hold the microbial mat of lichen together.

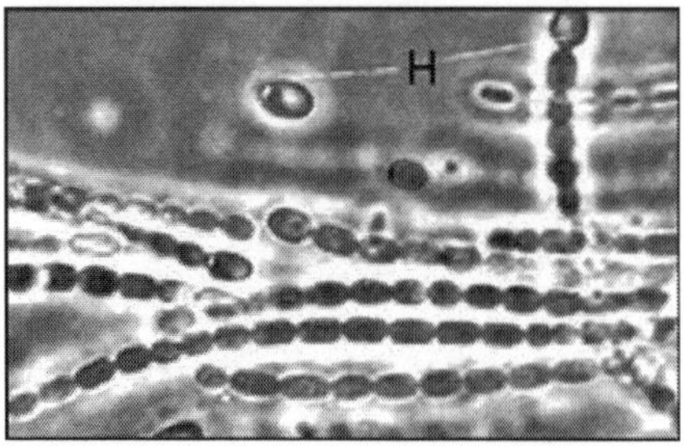

The cyanobacteria called Nostoc helps lichen produce food during photosynthesis.

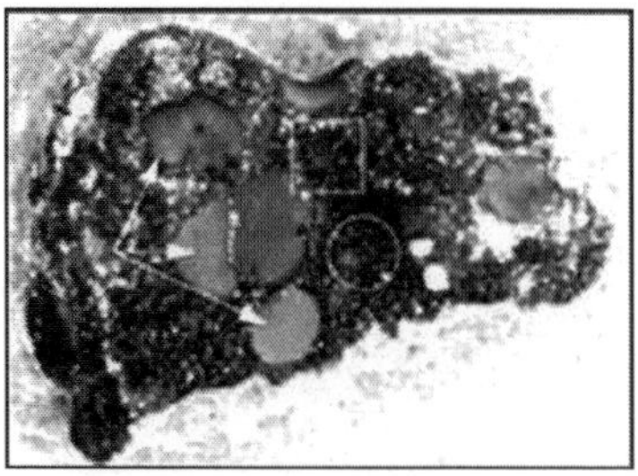

Fig. Microbial Crust

The microbial crust found in the desert is all dried up for most of the year. All it takes is a little bit of water to make it active again.

This is the microbial crust from the picture to the left after it was put in water. The arrows are pointing to a kind of lichen in this section, another form of lichen is inside the square, and cyanobacgteria are inside the circle.

AIRBORNE MICROBES

Airborne microbes cause a lot of illnesses and diseases in humans. Microorganisms can enter the air when a human or animal sneezes, or by the wind picking up the light particles and blowing them where humans are. When a human sneezes microorganisms leave the lungs at around 200 miles per hour. Some of the microorganisms that are growing in the mucus in the respiratory tract enter the air with the moisture particles that are sneezed out of the lungs. These microorganisms can be breathed into the lungs of another person and that person could get sick.

HOW ARE MICROORGANISMS IDENTIFIED

Microorganisms are put into groups, but a lot of microorganisms can belong to more than one group. One way that microorganisms are grouped is by the temperature in their environment. Another way to organize microorganisms is by placing them in either the prokaryot or eukaryot group.

HOW DO MICROORGANISMS REPRODUCE

Thermophiles reproduce either by sexual or asexual reproduction. Sexual reproduction requires a male and female organism, but asexual reproduction happens by cell division, mitosis. Thermophilic fungi reproduce by producing male and female spores that come in contact with each other to produce a new organism.

WHAT DO MICROORGANISMS DO

Microorganisms also are responsible for building fertile soil for plants to grow in. Microbes stick to the roots of plants and decompose dead organic matter into food for the plant to absorb. The plants that live and grow because of the microorganisms that live on them make a home for other animals to live in. Some microorganisms make people, animals, and plants sick, but others make people well and kill the bacteria on plants that make them sick. Drug companies that make medicines use hundreds of different microorganisms to make medicines that will help cure diseases. Human waste products are broken down into safer particles by some microorganism. Scientists are always looking for new ways to use microbes, and only a few uses have been listed here.

HISTORY

EVOLUTION

Single-celled microorganisms were the first forms of life to develop on Earth, approximately 3–4 billion years ago. Further evolution was slow, and for about 3 billion years in the Precambrian eon, all organisms were microscopic. So, for most of the history of life on Earth the only forms of life were microorganisms. Bacteria, algae and fungi have been identified in amber that is 220 million years old, which shows that the morphology of microorganisms has changed little since the Triassic period. Most microorganisms can reproduce rapidly and microbes such as bacteria can also freely exchange genes by conjugation, transformation and transduction between widely-divergent species. This horizontal gene transfer, coupled with a high mutation rate and many other means of genetic variation, allows microorganisms to swiftly evolve to survive in new environments and respond to environmental stresses. This rapid evolution is important in medicine, as it has led to the recent development of 'super-bugs'—pathogenic bacteria that are resistant to modern antibiotics.

Pre-microbiology

The possibility that microorganisms exist was discussed for many centuries before their actual discovery in the 17th century. The existence of unseen microbiological life was postulated by Jainism which is based on Mahavira's teachings as early as 6th century BCE. Paul Dundas notes that Mahavira asserted existence of unseen microbiological creatures living in earth, water, air and fire.

Jain scriptures also describe nigodas which are sub-microscopic creatures living in large clusters and having a very short life and are said to pervade each and every part of universe, even in tissues of plants and flesh of animals. However, the earliest known idea to indicate the possibility of diseases spreading by yet unseen organisms was that of the Roman scholar Marcus Terentius Varro in a 1st century BC book titled On Agriculture in which he warns against locating a homestead near swamps:

- ...and because there are bred certain minute creatures which cannot be seen by the eyes, which float in the air and enter the body through the mouth and nose and there cause serious diseases.

In The Canon of Medicine, Avicenna stated that bodily secretion is contaminated by foul foreign earthly bodies before being infected. He also hypothesized that tuberculosis and other diseases might be contagious, i.e. that they were infectious diseases, and used quarantine to limit their spread. When the Black Death bubonic plague reached Andalusia in Spain, in the 14th century, Ibn Khatima wrote that infectious diseases were caused by contagious "minute bodies" that enter the human body. Later, in 1546, Girolamo Fracastoro proposed that epidemic diseases were caused by transferable seedlike entities that could transmit infection by direct or indirect contact, or even without contact over long distances. All these early claims about the existence of microorganisms were speculative and were not based on any data or science. Microorganisms were neither proven, observed, nor correctly and accurately described until the 17th century. The reason for this was that all these early studies lacked the microscope.

History of Microorganisms' Discovery

Anton van Leeuwenhoek was one of the first people to observe microorganisms, using a microscope of his own design, and made one of the most important contributions to biology. Robert Hooke was the first to use a microscope to observe living things; his 1665 book Micrographia contained descriptions of plant cells. Before Leeuwenhoek's discovery of microorganisms in 1675, it had been a mystery why grapes could be turned into wine, milk into cheese, or why food would spoil. Leeuwenhoek did not make the connection between these processes and microorganisms, but using a microscope, he did establish that there were forms of life that were not visible to the naked eye. Leeuwenhoek's discovery, along with subsequent observations by Lazzaro Spallanzani and Louis Pasteur, ended the long-held belief that life spontaneously appeared from non-living substances during the process of spoilage.

Lazzaro Spallanzani found that boiling broth would sterilise it and kill any microorganisms in it. He also found that new microorganisms could only settle in a broth if the broth was exposed to the air. Louis Pasteur expanded upon Spallanzani's findings by exposing boiled broths to the air, in vessels that contained a filter to prevent all particles from passing through to the growth medium, and also in vessels with no filter at all, with air being admitted via a curved tube that would not allow dust particles to come in contact with the broth. By boiling the broth beforehand, Pasteur ensured that no microorganisms survived within the broths at the beginning of his experiment. Nothing grew in the broths in the course of Pasteur's experiment. This meant that the living organisms that grew in such broths came from outside, as spores on dust, rather than spontaneously generated within the broth. Thus, Pasteur dealt the death blow to the theory of spontaneous generation and supported germ theory.

In 1876, Robert Koch established that microbes can cause disease. He found that the blood of cattle who were infected with anthrax always had large numbers of Bacillus anthracis. Koch found that he could transmit anthrax from one animal to another by taking a small sample of blood from the infected animal and injecting it into a healthy one, and this caused the healthy animal to become sick. He also found that he could grow the bacteria in a nutrient broth, then inject it into a healthy animal, and cause illness.

Based on these experiments, he devised criteria for establishing a causal link between a microbe and a disease and these are now known as Koch's postulates. Although these postulates cannot be applied in all cases, they do retain historical importance to the development of scientific thought and are still being used today.

DIFFERENT TYPES OF MICROORGANISMS

The word micro-organism is a general term for a (very) small organism, so small that the use of a microscope is required to see details of its structure. The study of micro-organisms (also known as microbes) is called microbiology, and it is increasingly relevant in Biology. Specialists divide microbiology according to their interests, as follows: Fill in the names in the table below.

Main groups of micro-organisms	**Study called**	**Specialist called**
viruses(not really living organisms like the rest)	Virology	Virologist
Bacteria	Bacteriology	Bacteriologist
Fungi	Mycology	Mycologist
Possibly less important protozoa	Protozoology	Protozoologist
Algae	Algology (phycology)	Algologist (phycologist)

Each of these groups includes organisms which can be seen to be useful to man (especially in the context of biotechnology), and others which are harmful (mainly because of diseases - of animals and plants, or spoilage - of stored products, especially food). There are broad similarities in the way that these micro-organisms grow, but there are distinct differences in detail which must be appreciated.

From a classification point of view, these micro-organisms are now thought to merit separating from other more familiar living organisms (Plants and Animals), so they have been given Kingdoms of their own:

Group of micro-organism	Kingdom
Bacteria and Blue-green "algae"	Monera
Protozoa and Algae	Protoctists (Protists)
Fungi	Fungi(as a separate Kingdom, not a subdivision of the plants)

Viruses are categorised according to a different system, because they are so unlike the others.

MICRO-ORGANISMS CAUSING DISEASE

If a micro-organism has an adverse effect on another organism, e.g. causing a disease in Man, perhaps by getting inside and damaging its cells, or affecting it with a chemical substance which it produces, it is said to be pathogenic (adjective) or a pathogen (noun). It may also be described as the causative organism of that disease. Pathogens can be said to be parasitic because they live at the expense of the other organism - their "host". All viruses are pathogenic because they enter cells and cause adverse effects. Some of the worst diseases of Man are viral, i.e. caused by a virus. Some bacteria (ones which cause headlines!) are pathogenic, and a few cause quite serious diseases. Other bacteria seem to be useful to Man in their correct contexts, whilst other apparently fairly harmless bacteria may cause diseases in certain circumstances, e.g. old or young people. In addition, there may be different varieties or strains (especially of bacteria and viruses) which show different characteristics from the normal type or species. Similarly, there are examples of protozoans (protoctists) and fungi which are pathogenic.

WHY ARE MICRO-ORGANISMS |SUCH POWERFUL PATHOGENS

Size and Distribution

Being so small, many micro-organisms can fit into a small space, or spread out (thinly!) over a large area, although most are not able to move of their own accord.

Reproductive Potential

Because they can multiply rapidly, it only takes a few bacteria or viruses to cause an infection. Usually, micro-organisms reproduce asexually, so they can produce millions in a few hours.

Variability

With such large potential populations, new varieties can arise due to mutation, and characteristics like resistance to antibiotics can spread easily, even between unrelated species!

Insidiousness

They cannot be seen directly, and whilst reproducing and preparing to spread they can cause great damage to cells and internal working parts of organisms.

PROTOZOA

Protozoa (more than 30 000 species!!!) are eukaryotic cells; the cell is by itself an organism where various vital functions take place: They are grouped in 4 main families based on their mode of locomotion or their cell organisation.

The most well-know are the plasmodium type, reponsible for paludism, the trypanosoma type, agent of disease of sleep and transmit to man by the fly Tsé-Tsé, the trichomonas type, reponsible sexually transmitted diseases or the paramecium, usually present in fresh water.

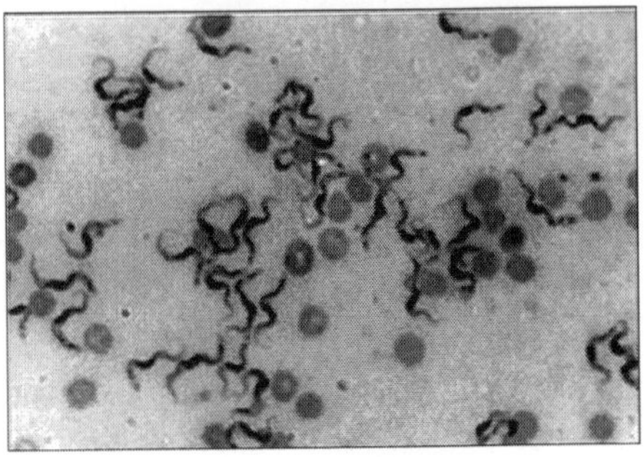

Fig. Photography of Trypanosoma

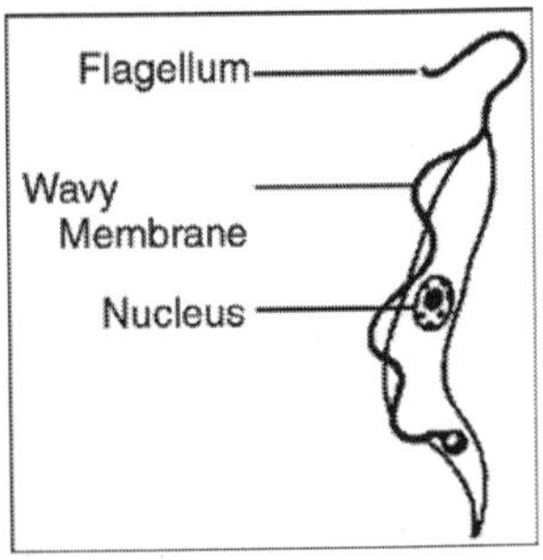

Fig. Trypanosoma Organization

MICROSCOPIC FUNGI

Only microscopic fungi are part of the micro-organisms: they are eukaryotic cells. The most significant group is yeasts' group, first micro-organism which has been used by the Man for the production of alcoholic drinks (since Antiquity). Pasteur's work on the yeast was the first step to the microbiology. Nowadays, those microorganisms are well-used in biomedical research and biotechnology. Fungi include 2 groups of "moulds" (yeasts are included in one), and also a group to which most of the mushrooms, toadstools, bracket fungi and puffballs belong. The best known evidence of fungi - their spore-bearing structures or fruiting bodies - are visible to the naked eye, but the main part of the fungal structure can only be seen with the aid of a lens or low power microscope.

From a structural point of view, there are two basic forms available to fungi:

1. Most moulds and higher fungi are filamentous forms - adapted for penetrating solid substrates,
2. Yeasts are examples of rounded cells - more suited to growth in liquids

Mucor ("the pin-mould") is often taken as an example of a simple filamentous form. This is named after the pin-like structures (sporangia) which project upwards, containing spores. If such a fungus spore lands on a suitable substrate (i.e. material capable of providing support and nutrition), it may germinate and produce a few filamentous strands (about 10 µm in diametre) called hyphae (singular - hypha) which are the feeding parts of the fungus and they branch repeatedly. A plant body (thallus) or colony thus builds up consisting of a fine network of hyphae, collectively called mycelium, which runs over the surface of the substrate, penetrating it to some extent.

Structure of Fungal Mycelium

Labels should include the following - but also read the notes which follow:

- Mycelium of branched hyphae with nuclei
- Cytoplasm
- Surface membrane (cell membrane) and
- Cell wall

There is no chlorophyll (colour whitish grey): photosynthesis is not carried out, and starch is not accumulated: oil or glycogen is the normal food store. The cell wallis made of chitin - a biological polymer containing N (and common with insects). A few fungi have cellulose cell walls. The structure is not divided into cells - it is called a coenocyte - and many minute nuclei are "loose" in the cytoplasm. Hyphae grow out radially to form a mat on a surface due to their need for oxygen, but the hyphae interweave in a complex fashion.

Spore Production (Sporulation)

Like many other fungi, Mucor produces spores asexually - there is no need for an encounter with another colony. An aerial hypha grows up vertically and the tip swells with cytoplasm containing many nuclei which form into many spores, each containing several nuclei. As the outer layer of the spore case (sporangium) breaks open, either dry spores are shot out and dispersed by air, or spores in a sticky substance are carried away by insects. Alternatively, Mucor can also produce sexual spores, as a result of one colony growing into close contact with another. Special hyphae swell and grow towards each other, then fuse together, and nuclei from each side fuse (fertilisation). The outer wall thickens so that a zygospore is produced, which is fairly long-lived and resistant to adverse conditions. After its dispersal, this will break open to produce a hypha which grows to support a sporangium, which once again releases spores.

Other examples of filamentous fungi:

Penicillium is a genus of filamentous fungi, most of which produce blue- green spores on distinctly shaped slightly different structures. This has many similarities withMucor, but different species have put put to various uses by Man. For example, Penicillium notatum and Penicillium chrysogenum produce penicillin which has been developed for medical use. Other species of Penicillium are responsible for the distinctive taste of blue cheeses, e.g. Stilton, Danish Blue, Roquefort, etc. Aspergillus is another genus of moulds, and one species is used by Man in the production of soy sauce. Fusarium is yet another genus of moulds which has been developed into a food (mycoprotein).

Structure of Yeast Cells Saccharomyces Spp

Yeast is a fungus which has been used by Man for a long time in the brewing and baking processes. Unlike most fungi, yeast cells are round or oval in shape. As they grow, yeast cells extend sections called buds which enlarge and eventually become detached, to lead a separate existence.

Yeast Growth Requirements

Like many fungi and bacteria, yeast may be grown on a semi-solid agar medium, but also in liquid media, in which it grows best. For a carbon source, yeast may use simple sugars, e.g. glucose, although some species can use other more complex carbohydrates, due to special enzymes they release.

This conversion involving fermentation by yeast is the basis of 2 major industries, based on techniques with a long history:

1. Brewing, where ethanol adds an undefinable extra value, acts as a preservative, and also extracts flavours from natural products, e.g. fruit juices (EEC specifies grapes) wine malted barley extract beer apple juice cider. This is also used as a feedstock for distillation industries spirits, e.g. brandy, whiskey, gin etc.
2. Baking, where CO_2 which is trapped in dough causes bread, etc, to "rise" and so produces a product with lighter texture when cooked.

Growth Rate of Micro-organisms

Under ideal conditions, bacteria reproduce extremely rapidly, perhaps doubling in numbers every 20 or 30 minutes. Fungi grow more slowly, but yeast population growth shows a similar situation. This exponential growth could theoretically result in incredibly large numbers within a few hours or days. In order to grow, micro-organisms mainly need a supply of suitable food, usually organic and inorganic, and an appropriate temperature.

CLASSIFICATION AND STRUCTURE

Microorganisms can be found almost anywhere in the taxonomic organization of life on the planet. Bacteria and archaea are almost always microscopic, while a number of eukaryotes are also microscopic, including most protists, some fungi, as well as some animals and plants. Viruses are generally regarded as not living and therefore are not microbes, although the field of microbiology also encompasses the study of viruses.

Prokaryotes

Prokaryotes are organisms that lack a cell nucleus and the other membrane bound organelles. They are almost always unicellular, although some species such as myxobacteria can aggregate into complex structures as part of their life cycle.Consisting of two domains, bacteria and archaea, the prokaryotes are the most diverse and abundant group of organisms on Earth and inhabit practically all environments where some liquid water is available and the temperature is below +140 °C. They are found in sea water, soil, air, animals' gastrointestinal tracts, hot springs and even deep beneath the Earth's crust in rocks. Practically all surfaces which have not been specially sterilized are covered by prokaryotes. The number of prokaryotes on Earth is estimated to be around five million trillion trillion, or 5 × 10, accounting for at least half the biomass on Earth.

Bacteria

Bacteria are practically all invisible to the naked eye, with a few extremely rare exceptions, such as Thiomargarita namibiensis. They lack membrane- bound organelles, and can function and reproduce as individual cells, but often aggregate in multicellular colonies. Their genome is usually a single loop of DNA, although they can also harbour small pieces of DNA called plasmids.

These plasmids can be transferred between cells through bacterial conjugation. Bacteria are surrounded by a cell wall, which provides strength and rigidity to their cells. They reproduce by binary fission or sometimes by budding, but do not undergo sexual reproduction. Some species form extraordinarily resilient spores, but for bacteria this is a mechanism for survival, not reproduction. Under optimal conditions bacteria can grow extremely rapidly and can double as quickly as every 10 minutes.

Archaea

Archaea are also single-celled organisms that lack nuclei. In the past, the differences between bacteria and archaea were not recognised and archaea were classified with bacteria as part of the kingdom Monera. However, in 1990 the microbiologist Carl Woese proposed the three-domain system that divided living things into bacteria, archaea and eukaryotes. Archaea differ from bacteria in both their genetics and biochemistry. For example, while bacterial cell membranes are made from phosphoglycerides with ester bonds, archaean membranes are made of ether lipids.

Archaea were originally described in extreme environments, such as hot springs, but have since been found in all types of habitats. Only now are scientists beginning to realise how common archaea are in the environment, with crenarchaeota being the most common form of life in the ocean, dominating ecosystems below 150 m in depth. These organisms are also common in soil and play a vital role in ammonia oxidation.

Eukaryotes

Most living things which are visible to the naked eye in their adult form are eukaryotes, including humans. However, a large number of eukaryotes are also microorganisms. Unlike bacteria and archaea, eukaryotes contain organelles such as the cell nucleus, the Golgi apparatus and mitochondria in their cells.

The nucleus is an organelle which houses the DNA that makes up a cell's genome. DNA itself is arranged in complex chromosomes. Mitochondria are organelles vital in metabolism as they are the site of the citric acid cycle and oxidative phosphorylation. They evolved from symbiotic bacteria and retain a remnant genome. Like bacteria, plant cells have cell walls, and contain organelles such as chloroplasts in addition to the organelles in other eukaryotes. Chloroplasts produce energy from light by photosynthesis, and were also originally symbiotic bacteria.

Unicellular eukaryotes are those eukaryotic organisms that consist of a single cell throughout their life cycle. This qualification is significant since most multicellular eukaryotes consist of a single cell called a zygote at the beginning of their life cycles. Microbial eukaryotes can be either haploid or diploid, and some organisms have multiple cell nuclei. However, not all microorganisms are unicellular as some microscopic eukaryotes are made from multiple cells. Of eukaryotic groups, the protists are most commonly unicellular and microscopic. This is a highly diverse group of organisms that are not easy to classify. Several algae species are multicellular protists, and slime molds have unique life cycles that involve switching between unicellular, colonial, and multicellular forms. The number of species of protozoa is uncertain, since we may have identified only a small proportion of the diversity in this group of organisms.

Animals

Mostly animals are multicellular, but some are too small to be seen by the naked eye. Microscopic arthropods include dust mites and spider mites. Microscopic crustaceans include copepods and the cladocera, while many nematodes are too small to be seen with the naked eye. Another particularly common group of microscopic animals are the rotifers, which are filter feeders that are usually found in fresh water. Micro-animals reproduce both sexually and asexually and may reach new habitats as eggs that survive harsh environments that would kill the adult animal. However, some simple animals, such as rotifers and nematodes, can dry out completely and remain dormant for long periods of time.

Fungi

The fungi have several unicellular species, such as baker's yeast and fission yeast. Some fungi, such as the pathogenic yeast Candida albicans, can undergo phenotypic switching and grow as single cells in some environments, and filamentous hyphae in others. Fungi reproduce both asexually, by budding or binary fission, as well by producing spores, which are called conidia when produced asexually, or basidiospores when produced sexually.

Plants

The green algae are a large group of photosynthetic eukaryotes that include many microscopic organisms. Although some green algae are classified as protists, others such as charophyta are classified with embryophyte plants, which are the most familiar group of land plants. Algae can grow as single cells, or in long chains of cells. The green algae include unicellular and colonial flagellates, usually but not always with two flagella per cell, as well as various colonial, coccoid, and filamentous forms. In the Charales, which are the algae most closely related to higher plants, cells differentiate into several distinct tissues within the organism. There are about 6000 species of green algae.

CONTROL OF MICROORGANISMS

Controling microorganisms can either be positive or negative:

- Positive control: You want to make them grow: Industrial Fermentations; beer, wine and bread making
- Negative control: You want to destroy them by:
    1. Physical or chemical means or

2. Antibiotics: Usually we mean negative control and the rest of this discussion relates to the destruction or inhibition of microbes:

PHYSICAL AND CHEMICAL METHODS

When disinfecting, one must be concerned with the most resistant life forms present. The following list of organisms is arranged according to ease (or difficulty) of destruction.

Most resistant to destruction (least sensitive to physical and chemical agents):

- Endospores
- Mycobacteria
- Fungal Spores
- Small Non-enveloped viruses (polio, rotavirus, rabies)
- Vegetative Fungal Cells
- Enveloped Viruses (Herpes, Hepatitis B, Hepatitis C, HIV)
- Vegetative Bacteria

Most Sensitive to destruction (least resistant to physical and chemical agents). When disinfecting, one must be concerned with interferring matter, such as organic matter (Pus, Blood, Feces, etc.) which inactivates certain disinfectants (halogens). Soap inactivates Quaternary Ammonium Compounds. When disinfecting, one must remember that "Things Take Time."

No disinfecting treatment is instantaneous.

- Heat: Denatures proteins; Moist heat is more effective than Dry heat.
  - Boiling: Vegetative cells are killed within minutes. Spores can survive.

What is Tyndallization?

- Steam under pressure: Autoclave. Standard load = 121oC (250oF), 15 lbs. steam pressure for 15 min. This kills spores, therefore this is sterilizing. Things can go wrong therefore it is important to have quality control.: SPORE TEST also there are colour change indicators.
- Pasteurize: Kills vegetative bacteria. Thermoduric bacteria will survive.
- Dry Heat - Baking - 160oC (320oF) for 2 hours, or 170oC (338oF) are common sterilizing treatments.
- Incinerate

- Filtration: See the discussion and figure in Tortora et. al. on this. Membrane filters (MilliporeTM) are used to sterilize heat sensitive liquids. HEPA filters are used to sterilize air in biohazard hoods.
- Ultrasonic cleaners: Good for removing organic contaminants - "presterilize." Not bacteriocidal.

RADIATION

Ultraviolet Light (See Tortora et.al., for a description of what UV light does to cells). UVB = 280-320 nm; UVA = 320-400 nm Germicidal lamps are used to disinfect air and surfaces. UV does not penetrate glass, plastic or water very well. UV light does harm eyes and skin. Ionizing Radiation = X-Rays and Gamma Rays - sporocidal and penetrate well.

- Chemicals: Many of these are called antiseptics or disinfectants or both.
  - Sterilizing Chemicals:
  - Ethylene Oxide
  - Beta-Propriolactone
  - Phenol (spores may withstand)
  - Halogens (iodine, chlorine, bromine)
  - Formaldehyde
  - Glutaraldehyde
- Phenol: (Carbolic Acid) Works well in organic materials therefore good for blood and body fluid decontamination. Neurotoxic. Has residual effect (lasts a long time). Tuberculocidal, Not Sporocidal.
- Alcohols: Ethyl and Isopropyl Tuberculocidal but not sporocidal. Kills enveloped viruses but not naked viruses. 70 per cent is commonly used concentration. Must make surfaces REALLY WET in order to achieve optimum killing.

HALOGENS

- Chlorine : Excellent disinfectant, sporocidal, however easily inactivated by organic material. Dilutions of household bleach between 1:10 - 1:100 are sporocidal after 10 minutes of treatment. 2-4 drops of bleach per litre of water can be used to treat drinking water (let stand 30 min.)
- Iodine: Similar to chlorine. Often found as a tincture or alcohol solution. Also found as an iodophor (betadineTM)where the iodine is complexed with an organic molecule. Iodophors are more stable than tinctures and they release iodine more slowly and steadily. Therefore they are less harmful to human tissue.
- Cationic Detergents: (Quaternary Ammonium Compounds) Benzalkonium chloride and Cetylpyridium chloride are some common QUATS found in mouth washes and sore throat

remedies. These are weak antiseptics which are active against a wide variety of vegetative bacteria. They are not tuberculocidal. They are not sporocidal. They do not kill Pseudomonas. Many QUATS are inactivated by soaps and other detergents.

- Alkylating agents: These are very powerful disinfectants. They are sporocidal and therefore sterilizing. Formaldehyde and glutaraldehyde are used to fix and preserve infectious tissues. These are also known to be carcinogens. Ethylene oxide is used in gas sterilizers. This is a very noxious, explosive gas.
- Heavy metals: Compounds made with mercury, silver, copper and tin have long been used for their antiseptic properties. In ancient times copper and silver containers were used to preserve water. Merthiolate, mercurochrome, silver nitrate and copper sulfate are all used as antiseptics. Silver nitrate has long been used to prevent neo-natal gonorrheal opthalmia and in the treatment of burns.
- Chlorhexidine: Active against vegetative bacteria and commonly used as a surgical scrub. This disinfectant is not sporocidal nor tuberculocidal.
- Hydrogen peroxide: This is a strong oxidizing agent which is very effective against vegetative bacteria on inanimate surfaces. The catalase in human tissue neutralizes H2O2 however the generation of oxygen bubbles helps to clean out wounds and is strongly inhibitory to anaerobic bacteria.

Table. Chemical Compounds Commonly Used as Antiseptics and Disinfectants

Compound	Type of Action	Applications
Hydrogen peroxide (3%)	Disinfectant/antiseptic	External surfaces, live tissue
Hypochlorites (0.5%) (Chlorox)	Disinfectant	External surfaces, non-living
Iodine (1% in 70% alcohol)	Disinfectant/antiseptic	External surfaces, live tissue
Iodophors (70 ppm available I2)	Disinfectant/antiseptic	External surfaces, live tissue,
Lysol (5%)	Disinfectant	External surfaces, non-living
Phenol (5%)	Disinfectant	External surfaces, non-living

Hexachlorophene	Disinfectant/antiseptic	External surfaces, live tissue
(pHisohex)		surgical scrub
Formaldehyde (4%)	Disinfectant	External surfaces, non-living
Zephrin and other compounds	Disinfectant	Exernal surfaces, non-living
Alcohol (ethyl or isopropyl at 70%)	Disinfectant/antiseptic	External surfaces, unbroken
Organic mercury (merthiolate, mercurochrome)	Disinfectant/antiseptic	External surfaces
Potassium permanganate	Antiseptic	Superficial skin fungus
Silver nitrate (1%)	Antiseptic	Prevent newborn eye infections
Ethylene oxide gas (12%)	Sterilizing disinfectant	Linens, heat labile plastics
Glutaraldehyde	Sterilizing disinfectant	Metal instrument sterilization
Formaldehyde (20% alcohol)	Sterilizing disinfectant	Metal instrument sterilization in

CHEMOTHERAPEUTIC AGENTS - THE ANTIBIOTICS

Most are derived from compounds which are biosynthesized by other microorganisms. Streptomyces and Penicillium are two organisms which have given us antibiotics.

The famous immunologist Paul Ehrlich devoted much of his career looking for the "Magic Bullet" or the chemical compound with selective toxicity. That is, such a compound would be toxic for the infecting microbe but not for the human host. This dream was achieved by Flemming and Florey with the discovery and mass production of penicillin.

Antibiotics achieve selective toxicity by exploiting the differences between eucaryotes and procaryotes. The most profound differences are at the cell wall and at the ribosome level. It follows then that most of our antibiotics either inhibit bacterial cell wall synthesis or inhibit the synthesis of bacterial proteins. For instance: ampicillin and penicillin interfere with cell wall synthesis; chloramphenicol, gentamicin, streptomycin and tetracycline interfere with protein synthesis.

How does triple sulfa work?

Antibiotic resistance is a growing problem which has at its source the indiscriminate and inappropriate use of antibiotics. Some of the current "Super Bugs" include methicillin resistant Staphylococcus aureus (MRSA), vancomycin resistant enterococcus (VRE) and multidrug resistant Mycobacterium tuberculosis. It is important that physicians order sensitivities on bacteria isolated from infection sites so that appropriate antibiotics can be prescribed.

Also, it is important that the infection be hit hard with high, optimum doses of the antibiotic and that tissue levels of the antibiotic remain high throughout therapy. Therapy must continue until all of the pathogens are dead this means that the patient must continue to take the antibiotic even after the symptoms are gone.

Classes of Antibiotics

- Antibiotics that inhibit cell wall synthesis
- Beta-Lactams
  a. Examples : Penicillins, cephalosporins, carbapenems, monobactams
  b. Usually bacteriocidal
  c. Inhibit the last step in peptidoglycansynthesis (transpeptidation)
  d. May trigger endogenous enzymes that degrade eptidoglycan
  e. Spectrum of various beta-lactams varies
- Glycopeptides
  a. Examples: Vancomycin, teichoplanin
  b. Prevents peptidoglycan cross-linking
  c. More effective against gram positive bacteria than gram negative bacteria
  d. Important in treating multi-drug resistant bacteria
- Phosphonomycin
  a. Inhibits synthesis of peptidoglycan
- Bacitracin
  a. Inhibits synthesis of peptidoglycan
  b. Used topically because of toxicity
- Antibiotics that inhibit protein synthesis
- Aminoglycosides
  a. Examples : Kanamycin, gentamycin, streptomycin, tobramycin, amikacin, netilmicin, neo-mycin, spectinomycin
  b. Bind 30s ribosomal subunit and prevent protein synthesis
  c. Bacteriocidal
  d. Potentially nephrotoxic and ototoxic

- Tetracylines
  a. Examples: Tetracycline, doxycycline
  b. Bind 30s ribosomal subunit and prevent protein synthesis
  c. Generally bacteriostatic
  d. Avoid use in pregnancy and in children under 8 because of interference with bone development and brown staining of teeth
- Chloramphenicol
  a. Example: Chloramphenicol
  b. Binds 50s ribosomal subunit and prevents protein synthesis. May also interfere with mitochondrial ribosomes leading to toxicity problems.
  c. Used to treat typhoid fever
- Macrolides
  a. Examples: Erythromycin, azithromycin
  b. Bind 50s ribosomal subunit and prevents the elongation of peptide chain
  c. Bacteriostatic for most bacteria; bacteriocidal for some gram-positive bacteria
- Lincosamides
  a. Examples: Lincomycin, clindamycin
  b. Bind 50s ribosomal subunit and prevents elongation of the peptide chain
  c. Bacteriostatic for most bacteria; bacteriocidal for some gram- positive bacteria
  d. Clindamycin penetrates bone well and also hits many gram positive and gram negative anaerobes — except that Clostridium difficile is often resistant and this may lead to pseudomembranous colitis
- Streptogramins (ketolides)
  a. Example: Synercid (mixture of dalfopristin and quinupristin)
  b. Binds 50s ribosomal subunit and prevents elongation of the peptide chain
  c. Individual components are bacteriostatic, the mixture is bacteriocidal
  d. Used for antibiotic resistant strains of Enterococcus and Staphylococcus

• Antibiotics that Inhibit Nucleic Acid Synthesis or DNA Replication

- Quinolones
  a. Examples: Ciprofloxacin, norfloxacin, naladixic acid
  b. Bind DNA gyrase and prevent DNA supercoiling

c. Broad bacteriocidal activity; effective against some intracellular bacteria

- Doesn't hit Streptococcus nor anaerobes very hard, therefore tends to be benign to the oral an the intestinal flora
- Metronidazole
  a. Example: Flagyl
  b. Interferes with DNA replication and creates breaks in the DNA
  c. Must be activated by flavodoxin or ferredoxin of the host cell — these are found in anaerobic bacteria.
  d. Effective against anaerobic and microaerophilic bacteria.

Also effective against some protozoa such as Trichomonas, Giardia, and Entamoeba.

- Rifampin
  a. Example: Rifadin
  b. Prevents the synthesis of bacteria messenger RNA
  c. Broad spectrum
  d. Used in tuberculosis therapy and also to treat the contacts of bacterial meningitis

• Antibiotics that Interfere with Metabolic Pathways

- Trimethoprim and sulfonomides
  a. Examples: Bactrim, Septra
  b. Inhibit enzymes in the tetrahydofolate biosynthetic pathway
  c. Broad spectrum against bacteria, some fungi (Pneumocystis carinii), and protozoa

MICROORGANISMS AND HUMAN BEINGS

The first words which come to mind when one speaks about micro-organisms are " germs ", " diseases "... Obviously, all the micro-organisms are not responsible for diseases (remember only 3 per cent of the bacteria are pathogenic), and can even be useful for the man: they are used in fermentations for alcoholic drinks or for research matters.

DISEASES

In this section, we will show how diseases and microorganisms are directly related.

Protozoa

Among the 30 000 species of protozoa which exist, the Trypanosome kind is one of most known because of the disease it causes. It is transmitted by the Tsé-Tsé fly which contaminates itself by aspiring the blood of a sick man. The disease of the sleep is exclusively African. The infected man has an intermittent fever, oedemas, a hypertrophy of lymphatic ganglia, nervous disorders (agitation or somnolence). Death is the usual end of the disease. If it is not treated, the disease lasts several years. This scary disease was stopped thanks to chemotherapy and nowadays, it is not a social plague anymore.

Fungi

In a sick immune system, germs such as Candida fungi (responsible for mycose) usually harmless may develop themselves and be responsible for diseases: Lily of the valley, the genital mucous membranes (vulvo-vaginits), the folds (intertrigo) on the level of the groin or between the toes. The dermatophytes are responsible for intertrigo and several varieties of tineas (patches with loss of hair). Some can be located in the lung as well: aspergillosis, cryptococcosis, histoplasmosis. There are many medicine like antifongic, different from antibiotics, used especially locally, against the cutaneous mycoses.

Bacteria

There are about 200 bacteria species which are pathogenic for human beings, for example the tetanus, the leprosy,the cholera, the plague and the syphilis which are responsible for the most dealdly diseases this past century. The level of bacteria infection depends on the bacteria colony and on the health on the host (here, we consider the human body).

The ways to fight against bacteria infection are well known: antibiotics synthetized by other microorganisms (penicillium is produced by penicillin, main component of aspirin). Anyway, those medicines are way to much used and bacteria change to fight against them.

Viruses

The influenza is one of the diseases which medicine still cannot cure perfectly. The only medicine commonly known is to go to bed. Indeed this disease has been known for a very long time and still does not have a 100 per cent effective medicine. This disease is caused by the flu virus which appears under different forms. The flu virus causes fevers and loose coughs. Influenza, or pandemias Epidemics have already killed many people: in 1918, a worldwide epidemic killed more than 20 million people! The 1968 (influenza of Hongkong) and 1977(Russian influenza) epidemics were as deadly. Currently, the influenza kills 4 000 people per years in a country like France, and 20 000 during epidemics, affecting the sensitive people (especially old or feverish people).

There are various way of fighting against the virus: the usual medicine which causes a drop in the fever or vaccines. Medicine research has improved significantly since the discovery of the virus in 1933: we know several colonies of flu virus, that does not facilitate the work of the researchers! Moreover, every year, the virus changes form: this is why people need to be vaccinate every year against the well-known flu virus.

A new medicine type developed by researchers will be launched on the market soon: Zanamivir and the GS 4104 (active molecules of the medicine) interact on the active site of an enzyme and prevent viral particles to move from one cell to another and to infect new cells.

Prions

The mad cow and Creutzfeldt-Jakob diseases which affect humans are spongiform encephalopathy. Infected cows lose their balance. Infected human beings show balanced problems first and then madness. The brain of the people affected by this disease has typical lesions: it is full of small holes and looks like a sponge (the name of the disease derives from this term).

INDUSTRIAL USES

We prefer to limit our development to French disease (our origins) and we hope to help readers to discover a new culture. First, let's present the Bread which is not a French speciality but which is so important to our French tradition. The manufacture of a bread starts with the baking of dough made up of flour (of corn or barley, according to the type of bread) and of water to which we add yeast of bakery to make it raise. This yeast is called Saccharomyces Cerevisiae,microscopic fungus which degrades sugar, alcohol and carbonic gas according to the fermentation process.

This yeast is also used in the wine and Champagne making process: starting from the grape gathered at the beginning of October, we obtain by pressing a juice or "moût". This juice is then sent in large fermentation tanks where it undergoes the first fermentation called "alcoholic fermentation" which requires both wild and cultivated yeasts (such as Saccharomyces).

Then intervenes the malolactic fermentation which makes it possible to decrease the acidity and the aggressiveness of the wine. After those two fermentations, the wine is finally processed by each wine maker who try to mix all the characteristics of the different grapes. The wine is then mixed in a tank with a sown liquor of selected colony. It is then bottled, the bottles are lying in the cellars while a new alcoholic fermentation occurs to give foam. The wine must remain at least a year in bottles for the ordinary vintages, three years for the special vintages.During the aging process, different flavours appear and lead to a high quality wine.

Another speciality of France is the cheese: there are more than 400 French cheeses. For example, the "camembert" was born during the French Revolution. Marie Harel, farmer from the little town Camembert (Normandy), created it and the making process of nowadays is very similar to the one used in the past. Its current manufacture remains near to the past one.

Milk is coagulated under the action of rennet: we obtain a gel which is drained off, draining is accompanied by an acidification due to the action of lactic bacteria introduced into milk at the same time of the addition of rennet. Then the curdled milk is covered with a thin layer of salt and "penicillium camemberti". The paste is subjected to a refining from four to five weeks which corresponds to a phase of enzymatic digestion: this phase is called enzymatic digestion where enzymes created by the existing microorganisms destroy the main elements of the curds. The main characteristics such as the aspect and texture change and the cheese develops savour and flavour characteristics.

MICROORGANISMS' DISCOVERY

Anton van Leeuwenhoek was the first person to observe microorganisms, using a microscope of his own design, thereby making him the first microbiologist. In doing so Leeuwenhoek would make one of the most important contributions to biology and open up the fields of microbiology and bacteriology. Prior to Leeuwenhoek's discovery of microorganisms in 1675, it had been a mystery as to why grapes could be turned into wine, milk into cheese, or why food would spoil.

Leeuwenhoek did not make the connection between these processes and microorganisms, but using a microscope, he did establish that there were forms of life that were not visible to the naked eye. Leeuwenhoek's discovery, along with subsequent observations by Lazzaro Spallanzani and Louis Pasteur, ended the long-held belief that life spontaneously appeared from non-living substances during the process of spoilage. Lazzaro Spallanzani found that microorganisms could only settle in a broth if the broth was exposed to the air.

He also found that boiling the broth would sterilise it and kill the microorganisms. Louis Pasteur expanded upon Spallanzani's findings by exposing boiled broths to the air, in vessels that contained a filter to prevent all particles from passing through to the growth medium, and also in vessels with no filter at all, with air being admitted via a curved tube that would not allow dust particles to come in contact with the broth. By boiling the broth beforehand, Pasteur ensured that no microorganisms survived within the broths at the beginning of his experiment. Nothing grew in the broths in the course of Pasteur's experiment. This meant that the living organisms that grew in such broths came from outside, as spores on dust, rather than spontaneously generated within the broth. Thus, Pasteur dealt the death blow to the theory of spontaneous generation and supported germ theory. In 1876, Robert Koch established that microbes can cause disease. He did this by finding that the blood of cattle who were infected with anthrax always had large numbers of Bacillus anthracis.

Koch also found that he could transmit anthrax from one animal to another by taking a small sample of blood from the infected animal and injecting it into a healthy one, causing the healthy animal to become sick. He also found that he could grow the bacteria in a nutrient broth, inject it into a healthy animal, and cause illness. Based upon these experiments, he devised criteria for establishing a causal link between a microbe and a disease in what are now known as Koch's postulates. Though these postulates cannot be applied in all cases, they do retain historical importance in the development of scientific thought and can still be used today.

ROLE OF MICRO-ORGANISMS

THE SULPHUR CYCLE

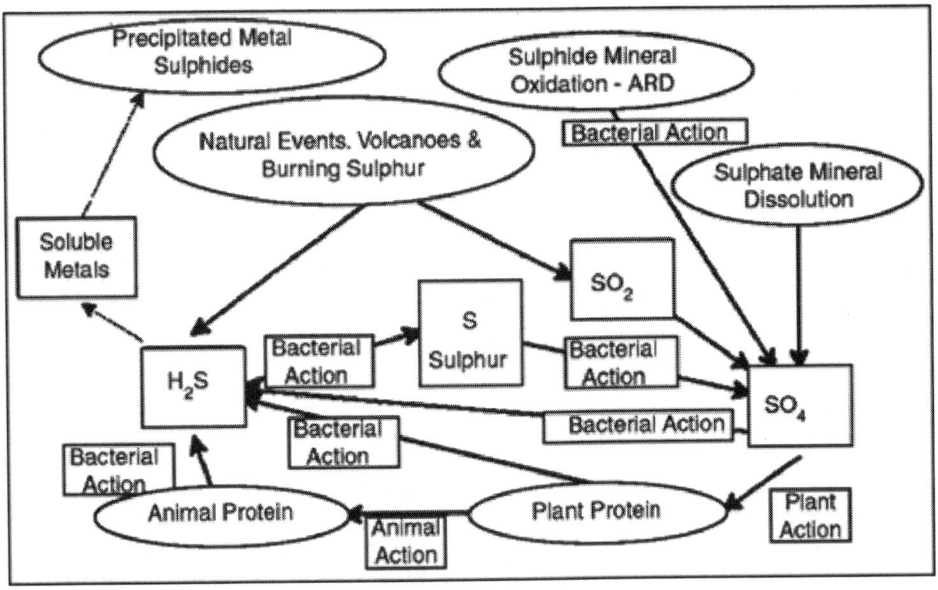

Fig. A Simplified Sulphur Cycle for Mineral Deposits

The environmental sulphur cycle involves many physical, chemical and biological agents.

A simplified schematic diagram of the cycle is shown above, which has been prepared to show the major phases of sulphur cycling with relation to mineral deposits. As such, the figure indicates the relationships between sulphur, S, hydrogen sulphide, H_2S, sulphur dioxide, SO_2, and the sulphate ion, SO_4—. In mineral form sulphur may be present as sulphides (e.g. pyrite, FeS_2, chalcopyrite, FeS.CuS, pyrrhotite, FeS) and/or sulphates (e.g. gypsum, $CaSO_4.2H_2O$, barite, $BaSO_4$).

Sulphur in minerals may move through the cycle as a result of the oxidation of sulphides to sulphate and/or the dissolution of sulphates. For example, oxidation of pyrite to sulphuric acid may be immediately followed, in situ, by acid neutralization by calcium carbonate (calcite) to form calcium sulphate (gypsum). The reaction of hydrogen sulphide with dissolved metal ions may precipitate metallic sulphides which are chemically indistinguishable from naturally occurring sulphide minerals. At some mines, sulphur is added to the cycle as sulphur dioxide in processes such as the Inco/SO_2 process for cyanide destruction in the treatment of tailings. This added sulphur is oxidized to sulphate ion, most of which remains free, but some of which combines with lime, CaO, in the tailings to form gypsum. For information on the sulphur cycle with respect to water quality monitoring.

THE ROLE OF MICRO-ORGANISMS IN THE SULPHUR CYCLE

Micro-organisms (most frequently bacteria) are often integrally involved in the chemical alteration of minerals. Minerals, or intermediate products of their decomposition, may be directly or indirectly necessary to their metabolism.

The dissolution of sulphide minerals under acidic conditions (ARD), the precipitation of minerals under anaerobic conditions, the adsorption of metals by bacteria or algae, and the formation and destruction of organometallic complexes are all examples of indirect micro-organism participation. Where minerals are available as soluble trace elements, serve as specific oxidizing substrates, or are electron donors/acceptors in oxidation-reduction reactions, they may be directly involved in cell metabolic activity.

There are three categories of oxidation-reduction reactions for minerals with micro- organisms:

- Oxidation by autotrophic (cell carbon from carbon dioxide) or mixotrophic (cell carbon from carbon dioxide or organic matter) organisms. Energy derived from the oxidation reaction is utilized in cell synthesis.
- Electron acceptance by minerals (reduction) for heterotrophic (cell carbon from organic matter) and mixotrophic bacteria. Chemical energy is used to create new cell material from an organic substrate.
- Electron donation by minerals (oxidation) for bacterial or algal photosynthesis (reaction is fuelled by photon energy).

NATURAL OXIDATION IN THE SULPHUR CYCLE

Oxidation of sulphur or sulphides for energy production is restricted to the bacterial genus Thiobacillus, the genus Thiomicrospira, and the genus Sulfolobus. These bacteria all produce sulphuric acid (i.e. hydrogen ions, H+, and sulphate ions, SO4—) as a metabolic product.

It is these bacteria that are known to accelerate the generation of Acid Rock Drainage (ARD) from pyritic and pyrrhotitic rocks under suitable conditions. report that sulphide oxidation catalysed by bacteria may have reaction rates six orders of magnitude (i.e. 1,000,000 times) greater than the same reactions in the absence of bacteria. Photomicrographs 1, 2 and 3, from, illustrate the shape and appearance of T. ferrooxidans: The bacteria develop flagella only if they are required for mobility in accessing energy sources.

Photomicrographs 1 (left), 2 (centre) and 3 (right): Thiobacilli from bacterial generator (no flagella) - left and centre - and grown on ferrous iron (flagella) - right. Magnification about x5000 (left), x20,000 (centre), x15000 (right) (from LeRoux, North and Wilson, 1973) ARD is the product formed by the atmospheric (i.e. by water, oxygen and carbon dioxide) oxidation of the relatively common iron-sulphur minerals pyrite and pyrrhotite in the presence of (catalysed by) bacteria (Thiobacillus ferrooxidans), and any other products generated as a consequence of these oxidation reactions. An important reaction involving T. ferrooxidans is the oxidation of ferrous to ferric iron (Fe++ to Fe+++) 4Fe++ + O_2 + 4H+ = Fe+++ + $2H_2O$ Ferric iron is a powerful oxidizing agent. Even at a Fe+++/Fe++ ratio of 1:1,000,000, a Redox potential of greater than +0.4 V is generated which is sufficient for the attack of most base metal sulphides.

The general equation for the ferric ion reaction with base metal sulphides is: MS + nFe+++ = Mn+ + S + nFe++ Consequently T. ferrooxidans, in generating Fe+++, is indirectly responsible for the dissolution of base metal sulphide minerals and the mobilization of metallic cations such as Cu++, Zn++, Pb++ and Cd++. Base metal sulphides react only very slowly with sulphuric acid in the absence of ferric iron.

The importance of Redox potential in determining metal solubility and transport can be clearly seen for copper in the Eh-pH diagram for the Cu- H_2O-O_2-S-CO_2 system.

The effects of bacteria upon the rate of dissolution of copper from chalcopyrite are highly pronounced,

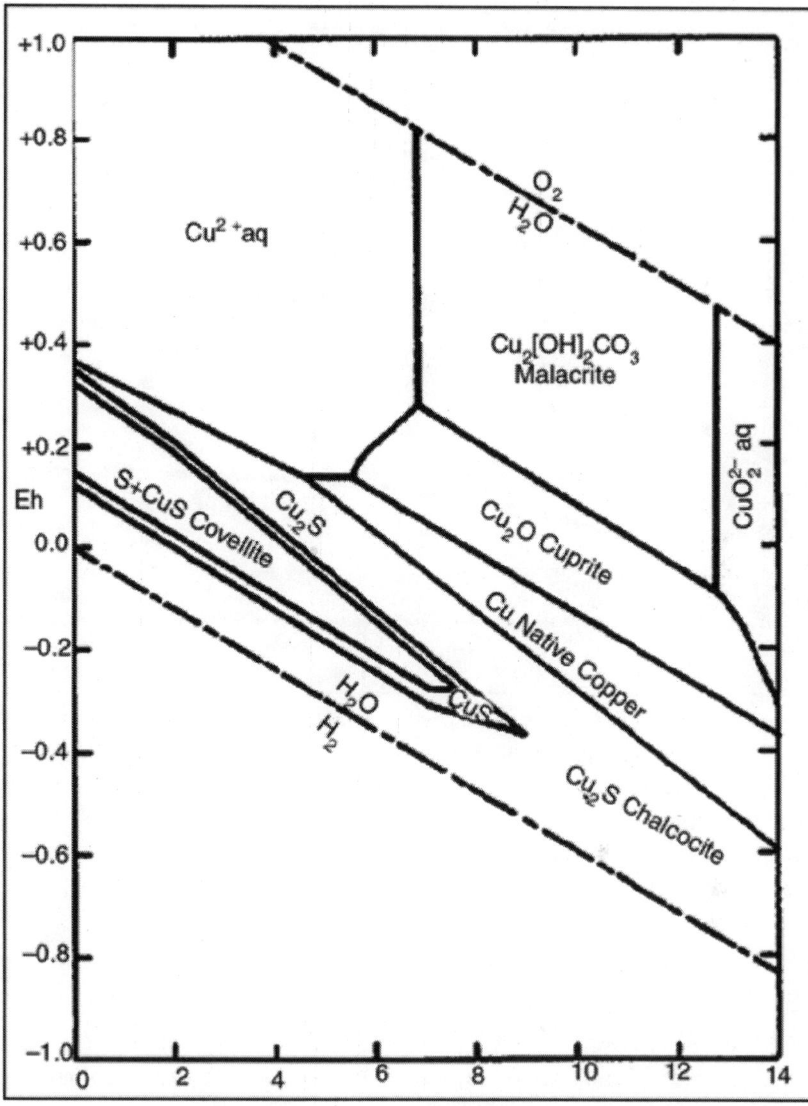

Fig. Eh-ph Diagram for Cu-H2O-O2-S-CO2 System (from Garrels and Christ, 1965)

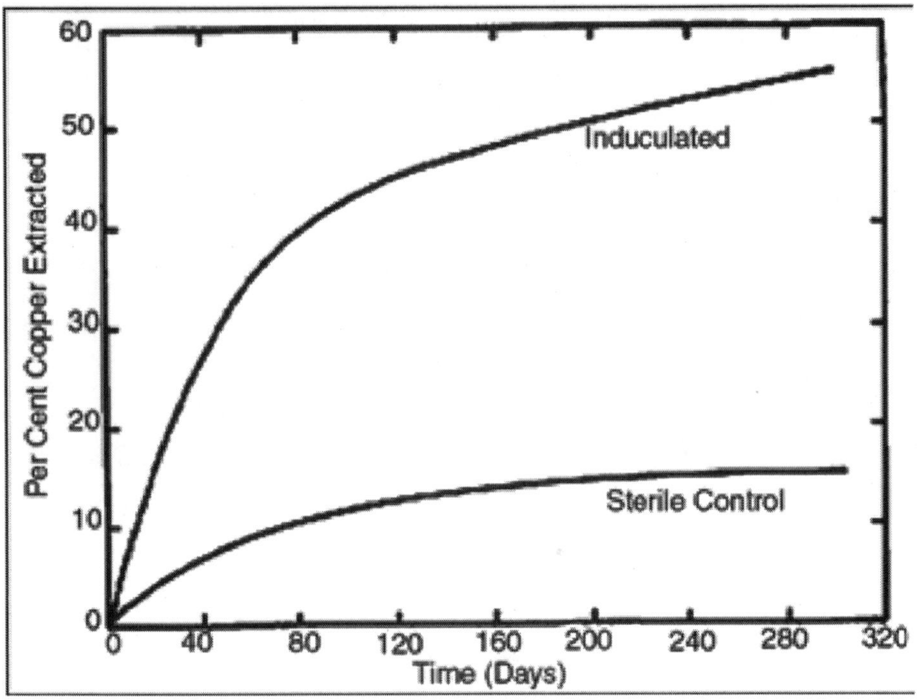

Fig. Effects of Bacteria upon the Rate of Disolution of Copper from Chalcopyrite (from Malouf and Prater, 1961) n general.

For substantial metal mobilization from base metal sulphides the following conditions must be met:

- Rric iron for sulphide oxidation
- T. ferrooxidans and oxygen for ferrous to ferric oxidation
- pH compatible with T. ferrooxidans habitat requirements, typically pH 1.5-3.5

The typical habitat pH of T. ferrooxidans of 1.5 to 3.5 is not one that develops spontaneously. It is currently believed that these conditions are produced by a consortium of bacteria acting in succession. Such a succession may include T. thioparus at neutral pH, giving way to dominance by metallogenium bacteria under mildly acid conditions (pH 3.5 to 4.5), and finally T. ferrooxidans dominance at low pH. The metabolic activity of T. ferrooxidans is temperature dependent, peaking at about 30-35 degrees Celsius, and falling with both increasing and decreasing temperature.have demonstrated that T ferrooxidans strains are site-specific.

From the above discussion it is clear that consideration of bacterial behaviour is most important in understanding the process of ARD generation.

This is particularly so when "kinetic" tests are used to predict the rate of generation of ARD in the field. Only if the bacterial conditions of testwork are identical to those in the field, can rates of ARD generation and/or metal solubilization be taken from laboratory kinetic testwork and used to predict field behaviour with any degree of confidence.

NATURAL REDUCTION IN THE SULPHUR CYCLE

The direct reduction of sulphate ions to hydrogen sulphide is effected in nature by specialized, strictly anaerobic bacteria of the genera Desulfovibrio and Desulfotomaculum.

These sulphate reducing bacteria (SRB) are heterotrophic (cell carbon from organic compounds) organisms that utilize sulphate, thiosulphate, S_2O_3–, sulphite, SO_3–, or other reducible sulphur-containing ions as terminal electron acceptors in their respiratory metabolism. In the process these sulphur-containing ions are reduced to hydrogen sulphide. The bacteria require an organic substrate which is usually a short chain acid such as lactic or pyruvic acid. In nature such substrates are generated by the fermentation activities of other anaerobic bacteria on more complex organic substrates.

Thus in natural systems, the specific requirement for a short chain acid by the SRB is met by the availability of a complex organic source and a mixed bacterial system.

Lactate is used by the SRB during anaerobic respiration to produce acetate according to the reaction:

$$2CH_3CHOHCOO- + SO_4- = 2CH_3COO- + 2HCO_3- + H_2S$$

This is the major natural process for the conversion (destruction) of sulphate ion. However, the process may be adapted to a controlled engineering process by the use of anaerobic reaction vessels and carbon monoxide, CO, and hydrogen, H_2, or partially oxidized propane or natural gas, as the energy source for the bacteria. This process of sulphate ion to hydrogen sulphide conversion is the first stage of a pilot plant that operated in 1996 at the former Britannia mine site in British Columbia, where the second stage utilized the hydrogen sulphide generated to precipitate copper (Cu++), Zinc (Zn++), and cadmium (Cd++) ions from ARD as metallic sulphides. Further information, including details of the performance of this pilot plant may be found have proposed the use of the underground mine at Faro, YT, as a giant underground SRB reactor to convert ARD sulphate ion to hydrogen sulphide, and precipitate zinc contained in the ARD as zinc sulphide. The proposed system would use the mixed bacterial system present in liquid cow manure as an SRB source, and sugar as a bacterial carbohydrate (energy) source.

SRB activity in natural wetlands is capable of metal sulphide precipitation from ARD as a result of sulphate to hydrogen sulphide reduction, and this concept may be extended to constructed wetlands. Dr. André Sobolewski treats this subject in considerable detail on his web site Wetlands for Treatment of Mine Drainage. Reviews of this subject have also been published by the Mine Environment Neutral Drainage (MEND) programme.

OTHER MICRO-ORGANISM REACTIONS IN THE SULPHUR CYCLE

Sulphate ion is taken up from soil by plants, which incorporate it into protein, and plant protein is consumed by animals that convert plant protein to animal protein. Death of plants and animals allows bacterial decomposition of protein in remains to produce hydrogen sulphide and other products, in processes involving many fungi, actinomycetes and bacteria such as the heterotroph Proteus vulgaris.

Some bacteria can function in the transition zone between aerobic and anaerobic environments. Hydrogen sulphide may be oxidized to sulphur by such bacteria which deposit elemental sulphur in their cells while using oxygen as the terminal electron acceptor.

Hydrogen sulphide may also be oxidized to sulphate photosynthetically by the bacteria, Chromtiacceae and Chlorobiaceae.

GROWTH REQUIREMENTS FOR MICROORGANISMS

A characteristic of microorganisms is their ability to grow and form a population of organisms. One of the results of microbial metabolism is an increase in the size of the cell. The many requirements for successful growth include those both chemical and physical.

CHEMICAL REQUIREMENTS

In order to grow successfully, microorganisms must have a supply of water as well as numerous other substances including mineral elements, growth factors, and gas, such as oxygen. Virtually all chemical substances in microorganisms contain carbon in some form, whether they be proteins, fats, carbohydrates, or lipids. Perhaps 50 per cent of a bacterium's dry weight is carbon. Carbon can be obtained from organic materials in the environment, or it may be derived from carbon dioxide. Both chemoautotrophic and photoautotrophic microorganisms obtain their energy and produce their nutrients from simple inorganic compounds such as carbon dioxide. Chemoauto-trophs do so through chemical reactions, while photoauto-trophs use photosynthesis. Among the other elements required by microorganisms are nitrogen and phosphorous.

Nitrogen is used for the synthesis of proteins, amino acids, DNA, and RNA. Bacteria that obtain nitrogen directly from the atmosphere are called nitrogen-fixing bacteria. They include species of Rhizobium and Azotobacter, both found in the soil. Phosphorus is an essential element for nucleic acid synthesis and for the construction of phospholipids. Oxygen is used by aerobic bacteria during the process of cellular respiration as a final electron acceptor.

For aerobic organisms, oxygen is an absolute requirement for their energy- yielding properties. Certain microorganisms grow in oxygen-free environments and are described as anaerobic. Organisms such as these produce odoriferous gases in their metabolism, including hydrogen sulfide gas and methane. Certain pathogenic species, such as Clostridium species, are anaerobic. Certain species of microorganisms are said to be facultative. These species grow in either the presence or absence of oxygen. Some bacteria species are microaerophilic, meaning that they grow in low concentrations of oxygen. In some cases, these organisms must have an environment rich in carbon dioxide. Organisms such as these are said to be capnophilic. Other chemical requirements for microbial growth include such trace elements as iron, copper, and zinc. These elements often are used for the synthesis of enzymes. Organic growth factors such as vitamins may also be required by certain bacteria. Amino acids, purines, and pyrimidines should also be available.

PHYSICAL REQUIREMENTS

Certain physical conditions affect the type and amount of microbial growth. For example, enzyme activity depends on the temperature of the environment, and microorganisms are classified in three groups just as to their temperature preferences: psychrophilic organisms (psychrophiles) prefer cold temperatures of about 0°C to 20°C; mesophilic organisms (mesophiles) prefer temperatures at 20°C to 40°C; thermophilic organisms (thermophiles) prefer temperatures higher than 40°C. A minimum and a maximum growth temperature range exist for each species. The temperature at which best growth occurs is the optimum growth temperature.

Another physical requirement is the extent of acidity or alkalinity, referred to as the pH of a solution. For most bacteria, the optimum pH is between 6.5 and 7.5. Since the pH of most human tissue is 7.0 to 7.2, these neutrophilic bacteria usually grow well in the body. Certain bacteria, such as those in sauerkraut and yogurt, prefer acidic environments of 6.0 or below.

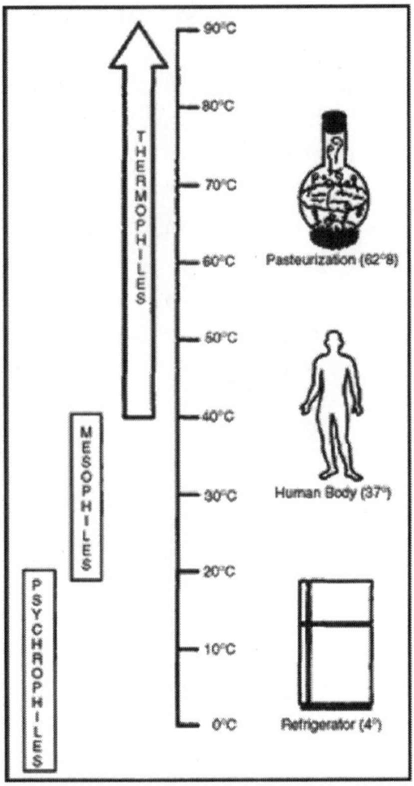

Fig. Three types of Bacteria and the Temperature Environments in which they thrive

These bacteria are said to be acidophilic. Molds and yeasts are among other common acidophilic microorganisms. Microbial growth proceeds best when the osmotic pressure is ideal. Normally, the salt concentration of microbial cytoplasm is about 1 per cent.

When the external environment also has a 1 per cent salt concentration, then the osmotic pressure is optimum. Should the external salt concentration rise, as when food is salted, water will flow out of the microbial cytoplasm by osmosis through the cell membrane into the environment, thereby causing the microorganisms to shrink and die.

By comparison, if exterior water is free of salt, it will flow through the cell membrane into the cytoplasm of the cell, causing the organism to swell and burst. Microorganisms that live in marine environments can tolerate high salt concentrations.

These organisms are said to be halophilic. They include diatoms and dinoflagellates, two types of unicellular algae that lie at the base of oceanic food chains. There are many other species of halophilic bacteria, fungi, protozoa, and algae.

MICROBIAL REPRODUCTION AND GROWTH

REPRODUCTION PATTERNS

During their growth cycles, microorganisms undergo reproduction many times, causing the numbers in the population to increase dramatically. In fungi, unicellular algae, and protozoa, reproduction involves a duplication of the nucleus through the asexual process of mitosis and a splitting of the cell in cytokinesis. Reproduction can also occur by a sexual process in which haploid nuclei unite to form a diploid cell having two sets of chromosomes. Various changes then follow to yield a sexually produced offspring. Sexual reproduction has the advantage of mixing chromosomes to obtain genetic variations not possible with asexual reproduction. However, fewer individuals normally result from sexual reproduction than from asexual reproduction. Bacteria reproduce by the asexual process of binary fission. In this process, the chromosomal DNA duplicates, after which the bacterial membrane and cell wall grow inward to meet one another and divide the cell in two.

The two cells separate and the process is complete. One of the remarkable attributes of bacteria is the relatively short generation time, the time required for a microbial population to double in numbers. The generation time varies among bacteria and often ranges between 30 minutes and three hours. Certain bacteria have very brief generation times. Escherichia coli, for example, has a generation time of about 20 minutes when it is dividing under optimal conditions.

THE GROWTH CURVE

The growth of a bacterial population can be expressed in various phases of a growth curve. The logarithms of the actual numbers in the population are plotted in the growth curve along the side axis, and the time is plotted at the base. Four phases of growth are recognized in the growth curve. In the first phase, called the lag phase, the population remains at the same number as the bacteria become accustomed to their new environment. Metabolic activity is taking place, and new cells are being produced to offset those that are dying. In the logarithmic phase, or log phase, bacterial growth occurs at its optimal level and the population doubles rapidly. This phase is represented by a straight line, and the population is at its metabolic peak. Research experiments are often performed at this time. During the next phase, the stationary phase, the reproduction of bacterial cells is offset by their death, and the population reaches a plateau.

The reasons for bacterial death include the accumulation of waste, the lack of nutrients, and the unfavourable environmental conditions that may have developed. If the conditions are not altered, the population will enter its decline, or death phase. The bacteria die off rapidly, the curve turns downward, and the last cell in the population soon dies.

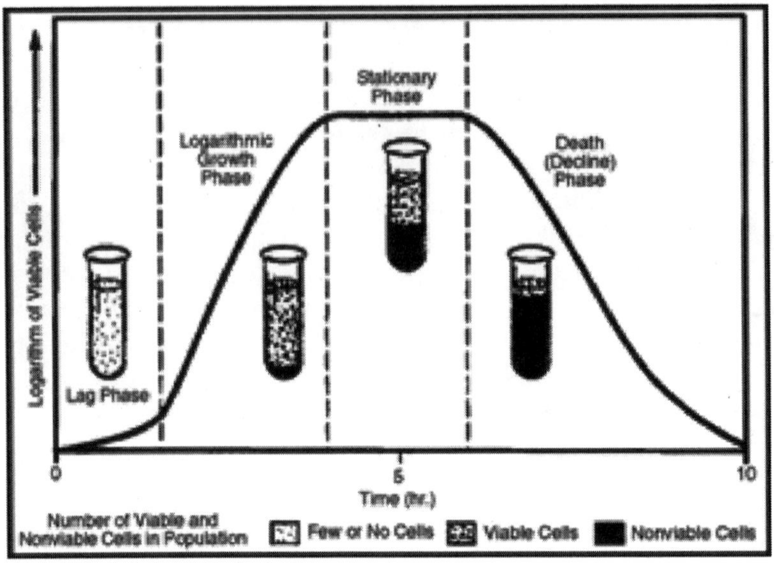

Fig. A Growth Curve of a Bacterial Population Showing the Four Major Phases of the Curve.

MICROBIAL MEASUREMENTS

In order to measure the number of bacteria in a population, various methods are available. In one method, known as the plate count method, a sample of bacteria is diluted in saline solution, distilled water, or other holding fluid. Samples of the dilutions are then placed in Petri dishes with a growth medium and set aside to incubate. Following incubation, the count of colonies is taken and multiplied by the dilution factor represented by that plate.

Generally, plates with between 30 and 300 colonies are selected for determining the final count, which is expressed as the number of bacteria per original ml of sample. Another measuring method is to determine the most probable number. This technique is often used to determine the number of bacteria in a sample of contaminated water.

Water samples are added to numerous tubes of single-strength and double-strength lactose broth. If coliform bacteria (such as E. coli) are present, they will ferment the lactose and produce gas. Judging by the number of tubes that contain gas at the end of the test, one may approximate the original number of bacteria in the water sample. Another evaluative method is by a direct microscopic count.

A specially designed counting chamber called a Petroff-Hausser counter is used. A measured sample of the bacterial suspension is placed on the counter, and the actual number of organisms is counted in one section of the chamber. The disadvantage of this method is that both live and dead bacteria are counted. Turbidity methods can also be used to assess bacterial growth.

As bacteria multiply in liquid media, they make the media cloudy. Placing the culture tube in a beam of light and noting the amount of light transmitted gives an idea of the turbidity of the culture and the relative number of bacteria it contains. The dry weight of a culture can also be used to determine microbial numbers. The liquid culture is dried out, and the amount of microbial mass is weighed on a scale.

It is also possible to measure the oxygen uptake of a culture of bacteria. If more oxygen is used by culture A than by culture B and all other things are equal, then it may be deduced that more microorganisms are present in culture A. A variation of this method called the biochemical oxygen demand (BOD) is used to measure the extent of contamination in a water sample.

PATHOGENICITY OF MICROORGANISMS

INTRODUCTION

The previous chapter mentioned that different types of microorganisms interact with human bodies on a regular basis. They can be harmless, harmful or beneficial. Harmful microorganims are also called pathogenic. This chapter looks at disease- causing bacteria and viruses.

COMMUNICABLE DISEASES

The ability of a microorganism to cause disease is called pathogenicity. There are several pathogens that can cause serious harm or even immediate death. Invasion and multiplication of pathogenic microorganisms in the body is called an infection. When we are infected by pathogens we become sick, which means that our bodies stop functioning properly. Infectious agents, such as bacteria, a virus, fungi or protozoa cause communicable diseases. Communicable diseases can be spread from one person to another.

INFECTION TRANSMISSION

All living organisms have a natural or acquired resistance mechanism called immunity. When we get sick, for example, we use different body cells and chemicals to fight bacteria. Bacteria in their turn use different chemicals to fight us. That is why infection is sometimes referred to as a race between pathogen and host organism. The infection can be transmitted by direct or indirect contact.

Direct Contact Transmission

Iinvolves any direct contact with an infected individual. Infection can be passed in water droplets through a sneeze, cough, laugh or exhalation and through bodily fluids. Most communicable diseases like colds, influenza, tuberculosis and HIV are spread from person to person through infected fluid droplets.

Indirect Contact Transmission

Is a method of spreading infection from person to person that involves contact with a contaminated object. Objects can become contaminated when touched by someone with an infection. The infected object is called a fomite. Another form of infection transmission through indirect contact is through the oral-feacal route, which usually involves ingesting contaminated water.

A third method of indirect contact involves vector-borne diseases, which are carried by animals and insects. A vector is an organism that serves as a 'go-between' in the transmission of a host-to-host disease.

Examples of diseases that can be transmitted via indirect contact are cholera, Salmonellosis and dysentery.

TYPES OF PATHOGENS

Natural or human-triggered changes in the environment might upset the natural balance between living organisms. These new environmental conditions may encourage pathogens, allowing them to multiply rapidly and increase the risk of exposing humans who share that environment. Here are the main groups of human pathogens with some examples of the diseases they cause.

EPIDEMIOLOGY

Epidemiology is the study of the patterns, causes and prevention of human diseases. An outbreak of a contagious disease that spreads rapidly and widely is called an epidemic. An epidemic occurring over a very large area is called a pandemic. The term 'epidemiologic triangle' is used to describe the intersection of host, agent and environment in analysing an outbreak. The prevalence of a disease is the number of diseased individuals at any one time (point prevalence) or over a given period (period prevalence). The incidence is the number of new cases of a disease that occur within a defined population over an established period of time.

HISTORY OF EPIDEMICS

The influenza (virus-caused) pandemic of 1918-1919 killed more people than World War I. It is considered the most devastating epidemic in recorded world history. More people died of influenza in a single year than in four years of the Black Death Bubonic Plague from 1347 to 1351. Bubonic plague is caused by bacteria called Yersinia pestis. These bacteria live in fleas on rats. Rats were abundant in medieval times due to very low hygiene standards, and the lack of sewage systems and hot, running water. Bubonic plague killed about a quarter of the European population of that time. Today this disease can be controlled by antibiotics but outbreaks sill occur sometimes.

PREVENTION OF INFECTIONS

The best way of fighting disease is to prevent it. The discovery and manufacture of antibiotics and vaccines in the early part of the 20th century have changed the state of people's health forever.

There are three stages of prevention of infections: primary, secondary and tertiary.

1. The Primary stage: Involves public education about infectious diseases.
2. The Secondary stage: Involves treating the actual infection that has already occurred by quarantining and/or vaccinating of infected individuals.
3. The Tertiary stage: Involves the recovery from illness.

BIOTERRORISM

Bioterrorism is the deliberate use of microorganisms, or their toxins, to cause death or disease. Biological and chemical agents that could be used include anthrax, small pox, West Nile virus and cholera. Biosecurity refers to the policies and measures taken for protecting a nation's food supply and agricultural resources from both accidental contamination and deliberate attacks of bioterrorism.

BENEFICIAL MICROORGANISMS

Previously, different types of microbes and the ways they reproduce were discussed. Some microbes are free-living organisms and others are parasites. Even though the words 'microbe' and 'bacteria' are associated with disease for most people, not all microorganisms are 'bad guys'. This topic looks at the interactions between microorganisms and the human body. Microorganisms can be harmless, beneficial or pathogenic, which means harmful. This chapter looks at the beneficial types of microorganisms.

What are Beneficial Microorganisms

Apparently, harmless and beneficial bacteria far outnumber the harmful varieties. Microbes are vital to the environment because they participate in the Earth's element cycles like the carbon and nitrogen cycles. Microorganisms are involved in the production of oxygen, biomass control and 'cleaning' the Earth of remnants of dead organisms.

Some microbes also lead a symbiotic type of lifestyle in most multicellular organisms. The community of beneficial microoraganims living in human intestines is called microflora.

Because microorganisms are capable of producing so many enzymes necessary for the building up and breaking down of organic compounds, bacteria are widely 'employed' by humans.

Nitrogen Cycle

Nitrogen is a very important chemical element of all living matter. It is an essential part of amino acids - the building blocks of proteins. Nitrogen in its gaseous form (N2) makes up 78 per cent of the atmosphere, but it cannot be absorbed and used as a nutrient by plants and animals.

It must be converted by nitrifying (nitrosomonas) bacteria, so that it can enter food chains as a part of the nitrogen cycle. The nitrogen cycle is the cyclic movement of nitrogen in different chemical forms from the environment to organisms and then back to the environment. The nitrogen cycle consists of several different processes: nitrogen fixation; ammonification (decay); nitrification; and denitrification.

The nitrogen cycle is also used in agricultural practices for soil enrichment.

Decomposition

Microbes 'clean up' waste products and remnants of dead organisms in a process called decomposition. The decomposition or stabilisation of organic matter by biological action is as old as life itself. The controlled microbial decomposition of organic matter is called composting.

The final product of composting is called compost. There are two types of composting:

1. Aerobic: With oxygen.
2. Anaerobic: Without oxygen.

In these processes, bacteria, fungi, moulds, protozoa and other saprophytic organisms feed upon decaying organic materials initially. In the later stages of decomposition, mites, millipedes, centipedes, springtails, beetles and earthworms further break down and enrich these composting materials.

Biotechnology

The industrial application of living organisms is called biotechnology. Humans have been using microorganisms for centuries. Today, biotechnology is a fast-developing industry.

Bioremediation

Bioremediation is the use of living organisms for cleaning up oil spills and soil and water pollutants. Sewage treatment techniques are based on biofiltration of some toxic organic material by converting it into something that can be safely discharged into the environment. Bacteria that break down environmental pollutants are sometimes called biofilters.

Pharmaceuticals

Some microbes are used for medicinal production. One of the most important groups of medicines, antibiotics, is produced by fungi and bacteria. The name antibiotics means 'against life'. It is appropriate, because they attack bacteria and other unicellular organisms that are pathogenic for humans. Most antibiotics used today were found originally in fungi. Fungi are saprophytes, meaning that they get their nourishment from dead animals or plant matter.

Microflora

Billions of bacteria live in the human digestive system. They form over a kilogram of our body weight. These bacteria are referred to as microflora, or gut flora. These bacteria break down food remains that have not been digested earlier in the digestive system. They stop harmful bacteria and fungus from invading the body. The 'gut flora' also produces vitamin K, which is essential for normal blood clotting.

Human microfolora consists of 400 different species of bacteria. Some of these are beneficial and others are potentially harmful. A balance between the two is vital for human health and wellbeing. One way of maintaining a balance between the beneficial and harmful bacteria in our intestines is to eat the types of food that contain beneficial bacteria. Beneficial bacteria that can be introduced into the digestive system through food are called probiotics. Most commercially-promoted fermented milk products with probiotic properties contain Lactobacillus bacteria or Bifidobacteria. Natural yogurts and Yakult, a fermented milk product, are examples of foods which contain probiotics.

Food industry

Fermentation is the process that produces alcoholic beverages or acidic dairy products. On a cellular level, fermentation is a way of obtaining energy without using oxygen. Fermentation involves the breaking down of complex organic substances into simpler ones.

Food like cheese, pickles, olives, sausages, chocolate, bread, wine, beer and soy sauce are all made with the help of different types of bacteria and yeast. In most of these food products, bacteria play a major role because they produce lactic acid.

❐

Chapter - 3

History of Plant Pathology

INTRODUCTION

Plant Pathology is defined as the study of the organisms and environmental conditions that cause disease in plants, the mechanisms by which this occurs, the interactions between these causal agents and the plant (effects on plant growth, yield and quality), and the methods of managing or controlling plant disease. It also interfaces knowledge from other scientific fields such as mycology, microbiology, virology, biochemistry, bioinformatics, etc.

Plant pathology is the scientific study of plant diseases caused by pathogens (infectious diseases) and environmental conditions (physiological factors). Organisms that cause infectious disease include fungi, oomycetes, bacteria, viruses, viroids, virus-like organisms, phytoplasmas, protozoa, nematodes and parasitic plants. Not included are ectoparasites like insects, mites, vertebrate, or other pests that affect plant health by consumption of plant tissues. Plant pathology also involves the study of pathogen identification, disease etiology, disease cycles, economic impact, plant disease epidemiology, plant disease resistance, how plant diseases affect humans and animals, pathosystem genetics, and management of plant diseases.

The study of plant pathology in New Zealand was begun in 1893 with the appointment of T.W. Kirk as Biologist to the newly formed Department of Agriculture. Apart from undertaking a considerable amount of work on the identification of diseases, Kirk did much to educate farmers and fruitgrowers to appreciate what diseases were and how they could best be controlled. In 1911 when Kirk became Director of the Horticulture Division, A.H. Cockayne was placed in charge of the Biological Laboratory, which was moved from Wellington in 1915 to the Department's Central Development Farm at Weraroa, near Levin. The development of the Laboratory was restricted by the First World War.

The post-war years saw the beginning of a new era in plant pathology in this country. G.H. Cunningham, who joined the staff in 1919 as mycologist, began a systematic survey of diseases attacking fruits, field crops, vegetables, and flowers. His work on the classification of fungi was also begun about this time. In 1925 he published the first New Zealand book on plant diseases – Fungous Diseases of Fruit Trees in New Zealand.

The Biological Laboratory returned in 1920 to Wellington where it remained until 1928 when the staff was transferred to Palmerston North to form the nucleus of the Plant Research Station, an organisation administered jointly by the Departments of Agriculture and of Scientific and Industrial Research. Within the new Station, Cunningham was placed in charge of a Mycological Laboratory. The next eight years was a period of considerable development and expansion in plant pathology. Investigations which until then had been mainly concerned with fungous and bacterial diseases were widened in scope to include virus and physiological diseases, techniques of spray application, and evaluation of the rapeutants used for disease control.

In 1936, following a reorganisation of agricultural research in New Zealand, the Department of Scientific and Industrial Research was made responsible for plant research and the Plant Research Station disbanded. As a result of the reorganisation, the Plant Diseases Division was established as a branch of the Department of Scientific and Industrial Research with G.H. Cunningham as Director. The Division moved in 1939 to permanent quarters at Mount Albert, Auckland, where modern laboratory and glasshouse facilities had been provided on an area of 16½ acres.

Plant Diseases Division, the centre of plant pathological work, has a permanent staff of 70 including 24 professional plant pathologists. Of the latter, 19 are stationed at Mount Albert where they are grouped in teams working in the fields of systematic mycology, economic mycology, bacteriology, virology, and therapeutant control of diseases, each group being responsible for both basic and applied research in its own specialist field. The remainder are stationed at substations, disease problems of field crops being studied by three professional officers at Lincoln in the South Island, while diseases of pasture plants are investigated by an officer at Palmerston North, and tobacco diseases by one at Nelson.

In addition to those employed by the Plant Diseases Division, there are about 12 other plant pathologists practising their profession. Within the Department of Scientific and Industrial Research, the Crop Research Division employs a pathologist to investigate virus diseases of potatoes, the

Fruit Research Division takes part in the investigation of fruit tree viruses, and a nematologist is attached to the Entomology Division. To assist with plant quarantine and advisory and extension work, the Department of Agriculture also employs several plant pathologists. In addition to their servicing work, these officers undertake research on diseases of vegetables and berry fruits and on chemical control of diseases. Research on diseases of forest trees is carried out by the plant pathologists employed by the Forest Research Institute of the State Forest Service. At Massey and Lincoln Agricultural Colleges, students are trained by qualified scientists who also undertake investigation of plant diseases. At least one plant pathologist is employed by a chemical- manufacturing firm. Scientists in other fields also contribute to work in plant pathology. Thus plant breeders at the Crop Research and Grasslands Divisions of the Department of Scientific and Industrial Research are concerned with the breeding of disease-resistant varieties of field crop, vegetable, and pasture plants, and chemists of the Dominion Laboratory assist with studies on therapeutant chemicals.

An extension and advisory service concerning plant diseases is provided by the Farm Advisory and Horticultural Divisions of the Department of Agriculture, whose officers make substantial contributions to plant pathology with their field observations and by carrying out plant-disease surveys and conducting field trials on disease control. They also assist pathologists by drawing to their attention new disease problems, collecting disease material for identification and study, and by assisting in the selection of disease-resistant varieties. The Department of Agriculture also provides a plant quarantine service and administers seed certification schemes which have as one of their aims the production and distribution of disease-free seed.

Introduction and use of control measures have been successful in greatly reducing losses caused by plant diseases, but in spite of this and the fact that hundreds of thousands of pounds are spent annually on disease control, direct losses greatly exceed £10,000,000 and are probably considerably in excess of £20,000,000 a year. Also of major importance are indirect losses brought about by the inability, because of disease, to grow economically certain crops, such as brassicas and cereals, in many parts of the country. At a moderate estimate the combined ravages of fungous, bacterial, and virus diseases reduce the value of our brassica crops, which are widely used as supplementary stock food, by £5,000,000 a year, while diseases of fruit crops take an annual toll of more than £1,000,000. In some seasons losses in cereal crops also amount to more than £1,000,000, and diseases in vegetables regularly cause losses of the order of millions of pounds a year. No crop grown in the Dominion is unaffected.

DEFENCE AGAINST PLANT DISEASES

The main concern of plant pathologists in New Zealand is to gain an understanding of our diseases and their significance in the country's economy and to minimise the losses they cause, generally in one or more of the following three ways: (1) By preventing the introduction of diseases not already here; (2) By eradicating those not fully established; and (3) By control of those which cause damage with chemical or cultural methods or by the development of resistant varieties.

Some serious plant diseases now confined to other countries, if established here, could cause losses amounting to millions of pounds annually. Hence plant quarantine regulations are enforced. Plants and seeds not known to carry diseases or pests which could be dangerous to our crops may be imported without restriction but are inspected at the port of entry. Others require an entry permit, must be grown for one season in a specified area and kept under observation by officers of the Department of Agriculture who certify them as disease-free before distribution. The introduction of plants of a third category is prohibited except in small numbers under special licence. These are grown under strict quarantine conditions at the Plant Diseases Division, Mount Albert, and released only after they have been declared free from disease.

Although eradication has been attempted on a number of occasions, it is rarely practicable because a disease is usually well established by the time it is recognised. An attempt to eradicate fire blight (Erwinia amylovora) failed but others have been gratifyingly successful. Citrus canker (Xanthomonas citri) has been eliminated from the large orchard areas at Kerikeri and Otumoetai, but there are still small pockets of infection in Taranaki, while an outbreak has recently been located in Auckland. Again, though onion yellow dwarf, a serious virus disease of onions, was well established by 1939 in the onion- growing area of Marshland, Christchurch, it has not been observed there for the last 20 years.

Regulations enforcing sanitation and prohibiting the growing of onion- seed crops, and alternative hosts brought about the eradication of this disease from the district. Even if not completely eliminated from the country, onion yellow dwarf has been successfully kept out of the major onion-growing areas. Attempts to eradicate Verticillium cinerescens of carnations, and the virus diseases Moorpark mottle of apricots and plum fruit crinkle appear to have been successful. Hot water treatment of barley, introduced during the 1920s, together with the production of mother seed and certification of seed crops, has eliminated the serious diseases of loose smut (Ustilago tritici) and covered smut (Ustilago hordei) from our malting barleys.

There are many ways of keeping diseases in check. Their incidence and severity may be greatly influenced by cultural factors such as soil type, date of sowing, size of plantings, crop rotation, sanitation, etc. Knowledge of these factors is used to reduce losses. Perhaps the most satisfactory way is to use resistant varieties, to the selection and breeding of which much attention has been given, with marked success in increased production of field crops and vegetables. Heavy losses can occur through the use of diseased stocks of vegetatively propagated plants. A seed potato certification scheme introduced in 1928 has done much to increase potato yields. Similarly, the health of fruit- tree nursery stock has been greatly improved by selection of disease-free root stocks and scion wood.

Work begun by New Zealand pathologists in the 1920s contributed much to knowledge of seed-borne diseases and their control. Since that date the general application of seed treatment to many field and vegetable crops has been of considerable benefit to the country's economy.

There are some serious soil-borne diseases particularly dangerous in glasshouses. Successful investigations on soil treatment have led to control of many of the more important of these, thereby achieving much heavier yields of tomatoes and other crops. Treatments have also been evolved and successfully applied for control of soil-borne organisms in the field.

Control of the majority of fungous and bacterial diseases is achieved by the application of therapeutant sprays. A great deal of research work has gone into finding the most effective material for specific diseases, the optimum concentrations, and the times and number of applications needed. As materials of high quality are essential for best results, the extent of control was greatly improved in 1938 by the introduction of a therapeutant certification scheme; but with increasing numbers of therapeutants becoming available it was not possible with limited scientific resources to maintain this. In 1959 it was replaced by an approval and registration scheme administered by an Agricultural Chemicals Board.

OBJECTIVES OF PLANT PATHOLOGY

Plant Pathology (Phytopathology) deals with the cause, etiology, resulting losses and control or management of the plant diseases. The objectives of the Plant Pathology are the study on:

- The living entities that cause diseases in plants;
- The non-living entities and the environmental conditions that cause disorders in plants; iii. the mechanisms by which the disease causing agents produce diseases;
- The interactions between the disease causing agents and host plant in relation to overall environment; and

- The method of preventing or management the diseases and reducing the losses/damages caused by diseases.

SCOPE OF PLANT PATHOLOGY

Plant pathology comprises with the basic knowledge and technologies of Botany, Plant Anatomy, Plant Physiology, Mycology, Bacteriology, Virology, Nematology, Genetics, Molecular Biology, Genetic Engineering, Biochemistry, Horticulture, Tissue Culture, Soil Science, Forestry, Physics, Chemistry, Meteorology, Statistics and many other branches of applied science.

CONCEPT OF PLANT DISEASE

The normal physiological functions of plants are disturbed when they are affected by pathogenic living organisms or by some environmental factors. Initially plants react to the disease causing agents, particularly in the site of infection. Later, the reaction becomes more widespread and histological changes take place. Such changes are expressed as different types of symptoms of the disease which can be visualized macroscopically. As a result of the disease, plant growth in reduced, deformed or even the plant dies.

CAUSES OF PLANT DISEASES

Plant diseases are caused by pathogens. Hence a pathogen is always associated with a disease. In other way, disease is a symptom caused by the invasion of a pathogen that is able to survive, perpetuate and spread. Further, the word "pathogen" can be broadly defined as any agent or factor that incites 'pathos or disease in an organism or host. In strict sense, the causes of plant diseases are grouped under following categories:

- Animate or biotic causes: Pathogens of living nature are categorized into the following groups.
  - Fungi
  - Bacteria
  - Phytoplasma
  - Rickettsia-like organisms
  - Algae
  - Phanerogams
  - Protozoa
  - Nematodes
- Mesobiotic causes: These disease incitants are neither living or non-living, e.g.
  - Viruses
  - Viroides

- Inanimate or abiotic causes: In true sense these factors cause damages (any reduction in the quality or quantity of yield or loss of revenue) to the plants rather than causing disease. The causes are:
  - Deficiencies or excess of nutrients (e.g. 'Khaira' disease of rice due to Zn deficiency)
  - Light
  - Moisture
  - Temperature
  - Air pollutants (e.g. black tip of mango)
  - Lack of oxygen (e.g. hollow and black heart of potato)
  - Toxicity of pesticides
  - Improper cultural practices
  - Abnormality in soil conditions (acidity, alkalinity)

When a plant is suffering, we call it diseased, i.e. it is at 'dis-ease'. Disease is a condition that occurs in consequence of abnormal changes in the form, physiology, integrity or behaviour of the plant. According to American Phytopathological Society, disease is a deviation from normal functioning of physiological processes of sufficient duration or intensity to cause disturbance or cessation of vital activities. The British Mycological Society defined the disease as a harmful deviation from the normal functioning of process. A plant is diseased when it is continuously disturbed by some causal agent that results in abnormal physiological process that disrupts the plants normal structure, growth, function or other activities. This interference with one or more plant's essential physiological or biochemical systems elicites characteristic pathological conditions or symptoms.

CLASSIFICATION OF PLANT DISEASE

To facilitate the study of plant diseases they are needed to be grouped in some orderly fashion. Plant diseases can be grouped in various ways based on the symptoms or signs (rust, smut, blight etc.), nature of infection (systemic or localized), habitat of the pathogens, mode of perpetuation and spread (soil-seed- and air-borne etc.), affected parts of the host (aerial, root disease etc.), types of the plants (cereals, pulses, oilseed, ornamental, vegetable, forest diseases etc.).

But the most useful classification has been made based on the type of pathogens that cause plant diseases. Since this type of classification indicates not only the cause of the disease, but also the knowledge and information that suggest the probable development and spread of disease alongwith their possible control measures. The classification is as follows:

- Infectious plant diseases:
  - Disease caused by parasitic organisms: The organisms included in animate or biotic causes can incite diseases in plants.
  - Diseases caused by viruses and viroids.
- Non-infectious or non-parasitic or physiological diseases: The factors included in inanimate or abiotic causes can incite such diseases in plants under a set of suitable environmental conditions.

PARASITISM AND PATHOGENESIS

An organism which lives in or on other living organisms and derives its nutrients from the latter is called parasite. The relationship between a parasite and its hosts is known as parasitism. Many fungi and most bacteria grow on a non-living substrate within a living plant. The organism of this type of mode of nutrition is called saprophyte. Based on the different types of modes or nature of nutrition, the relationship between the host and parasite or saprophyte is termed in many ways viz., obligate parasite (biotroph), obligate saprophyte, facultative parasite, facultative saprophyte, hemibiotroph and necrotroph (perthotrophs or perthophyte).

Parasitism in cultivated crops is common phenomenon. Any agent that can cause suffering or damage or disease is called a pathogen. In plant pathology, the term 'pathogen' is usually used to the living or infectious organisms. The ability of a pathogen or parasite to cause disease is known as pathogenicity. It is obvious that a plant becomes diseased when it is attacked by a pathogen or parasite. The ultimate condition i.e. disease occurs by passing through some distinct events. Thus, the genesis or chain of events or stages of disease development are called pathogenesis.

This is also called as disease cycle. The events that occur in specific order are namely inoculation, penetration, establishment of infection, invasion or colonization, growth and reproduction, dissemination and survival of the pathogen (over wintering or over summering in absence of the host).

The events will continue to repeat in the same order in presence of both the host and pathogen/parasite that may lead to severe disease condition.

Koch's Postulates

Robert Koch forwarded four essential procedural steps called postulates for correct diagnosis of a disease and its actual causal agent. The postulates are:

- Recognition: The pathogens must be found associated with the disease in the diseased plant. The symptom of the disease should be recorded.

- Isolation: The pathogen should be isolated, grown in pure culture in artificial media. The cultural characteristics of the pathogen should be noted.
- Inoculation: The pathogen of pure culture must be inoculated on healthy plant of same species/variety. It must be able to reproduce disease symptoms on the inoculated plant identical to step 1.
- Re-isolation: The pathogen must be isolated form the inoculated plant in culture media. Its cultural characteristics should be similar to those noted in step 2 (This step was added by E.F. Smith).

If all the postulates are proved true, then the isolated pathogen is identified as the actual causal organism responsible for the disease.

EFFECT OF PATHOGEN ON THE PLANTS

During the course of pathogenesis, normal activities of the infected host plant undergo malfunction. Consequently, morphological and physiological changes occur.

- Morphological or structural changes: Physiological malfunctioning of the host cells causes disturbances in chemical reaction which ultimately lead to some structural changes viz., overgrowth, phyllody, sterile flowers, hairy roots, witches broom, bunchy top, crown gall, root knot, leaf curling, rolling, puckering etc.
- Physiological changes:
  - Disintegration of the tissues by the enzymes of the pathogen.
  - Effect on the growth of the host plant due to growth regulators produced by the pathogen or by the host under the influence of the pathogen.
  - Effect on uptake and translocation of water and nutrients.
  - Abnormality in respiration of the host tissues due to disturbed permeability of cell membrane and enzyme system associated with respiration.
  - Impairing the phenomenon of photosynthesis due to loss of chlorophyll and destruction of leaf tissue.
  - Effect on the process of translation and transcription.
  - Overall reproduction system of the host.

SYMPTOMS OF PLANT DISEASES

A visible or detectable abnormality expressed on the plant as a result of disease or disorder is called symptom. The totality of symptoms is collectively called as syndrome while the pathogen or its parts or products seen on the affected parts of a host plant is called sign. Different types of disease symptoms are cited below:

- Necrosis: It indicates the death of cells, tissues and organs resulting from infection by pathogen. Necrotic symptoms include spots, blights, burn, canker, streaks, stripes, damping-off, rot etc.
- Wilt: Withering and drooping of a plant starting from some leaves to growing tip occurs suddenly or gradually. Wilting takes place due to blockage in the translocation system caused by the pathogen.
- Die-back: Drying of plant organs such as stem or branches which starts from the tip and progresses gradually towards the main stem or trunk is called die-back or wither tip.
- Mildew: White, grey or brown coloured superficial growth of the pathogen on the host surface is called mildew.
- Rusts: Numerous small pustules growing out through host epidermis which gives rusty (rust formation on iron) appearance of the affected parts.
- Smuts: Charcoal-like and black or purplish-black dust like masses developed on the affected plant parts, mostly on floral organs and inflorescens are called smut.
- Blotch: A large area of discolouration of a leaf, fruit etc. giving a blotchy appearance. White blisters: Numerous white coloured blister-like ruptures are surfaced on the host epidermis that forms powdery masses of spores of fungi. They are called white blisters or white rust.
- Colour change: It denotes conversion of green pigment of leaves into other colours mostly to yellow colour, in patches or covering the entire leaves.
- Etioliation: Yellowing due to lack of light,
- Chlorosis: Yellowing due to infection viruses, bacteria, fungi, low temperature lack of iron etc.
- Albino: Lack of any pigment and turned into white or bleached
- Chromosis: Red, purple or orange pigmentation due to physiological orders etc.
- Exudation: Such symptom is commonly found in bacterial diseases when masses of bacterial cells ooze out to the surface of affected plant parts and form some drops or smear, it is called exudation. This exudation forms a crust on the host surface after drying.
- Overgrowth: Excessive growth of the plant parts due to infection by pathogens. Overgrowth takes place by two processes:
    (i) Hyperplasia: abnormal increase in size due to excessively more cell division

(ii) Hypertrophy: abnormal increase in size or shape due to excessive enlargement of the size of cell of a particular tissue.

- Atrophy: It is known as hypoplasia or dwarfing which is resulted from the inhibition of growth due to reduction in cell division or cell size.
- Sclerotia: These are dark and hard structures of various shaped composed of dormant mycelia of some fungi. Sometimes, sclerotia are developed on the affected parts of the plant. Presence of sclerotia on the host surface is specifically called a sign of disease rather than symptom.

DEVELOPMENT OF EPIDEMICS

Sudden outbreak of a disease within a relatively short period covering a large area and affecting many individuals in a population is called epidemic. Although, this term was originally designated to the human diseases, now applied in the diseases of animals, poultry, plants etc. Epidemic form of plant diseases is called as epiphytotics.

For a disease to occur, coincidence of three parameters of disease triangle is essential, namely, the vulnerable host, virulent pathogen and favourable environment. Under such circumstances, the pathogen not only completes its life cycle but also undergoes repeated generations. Then an epidemic develops only when few repeated generations are completed by the pathogen on the same host. As each generation or cycle of the pathogen takes a few days for completion, the fourth parameter i.e. time factor (forms a disease tetrahedron or disease pyramid) is also involved in epidemic build up. In other words, epidemic growth is both temporal (pertaining to time) and a spatial (relating to space or area) process.

The initial stages of an epidemic growth curve have a lag phase, when the incubation period is longer, inoculum load is weak and prevalent environmental conditions are unfavourable. Subsequently, when the conducive conditions occur, the growth of the disease is rapid and the severity of the epidemic explodes like a time bomb.

Later, severity declines either due to unfavourable weather or crop maturity or both.

EMPIRICAL STUDIES OF SELECTION IN PLANT PATHOLOGY

Though it is very difficult to measure fitness and estimate selection coefficients, plant pathologists have made many attempts because of the profound implications of these measurements for the utilization of resistance genes. If the fitness cost of a mutation from avirulence to virulence is very low, then pathogen populations will retain the virulence mutation at a significant frequency even after the resistance gene is removed from the agroecosystem. In these cases, it is likely that the resistance gene will never again be effective. But if the fitness cost of the virulence mutation is high, then resistance genes are likely to remain durable, and it is possible that resistance genes can be recycled regionally or through time even after they have been overcome.

Most studies of pathogen fitness cannot be generalized because of the difficulties of conducting these experiments. Fitness should be determined at the level of populations rather than for individuals because a pathogen population is often composed of many different genotypes. But plant pathologists almost always work with just a few individuals to evaluate fitness to simplify the experimental methods. To measure fitness, sample sizes must be large when s (the selection coefficient) is small, but sample sizes are usually small (10-100 individuals) in field or greenhouse experiments due to practical limitations on the number of samples that can be handled. To further complicate the issue fitness can vary in different environments: the most fit isolate at 10°C may have the lowest fitness at 15°C. Fungi with mixed reproductive systems present special challenges when measuring fitness because selection can operate on alleles (sexual populations), genotypes (asexual populations) or simultaneously on both alleles and genotypes (mixed reproduction systems). While selection always occurs on the individual through its phenotype, it can be difficult to distinguish between selection for specific alleles (e.g. a virulence allele) within an individual and selection for the entire genotype (e.g. a set of virulence alleles in combination with fungicide resistance alleles) within an individual.

Different alleles can occur in a single genetic background as a result of mutation within clonal lineages. And the same allele can occur in different genetic backgrounds as a result of the same mutation in different clonal lineages.

GREENHOUSE EXPERIMENTS TO MEASURE PATHOGEN FITNESS

Many greenhouse experiments have been conducted to compare fitness of pathogen strains carrying different virulence alleles. The following representative examples were chosen to show how the experiments have been conducted and how the results have not led to a clear interpretation of the fitness cost associated with unneeded virulence alleles.

Kolmer (1993) used a diverse collection of Puccinia triticina (syn. Puccinia recondita f. sp. tritici which causes wheat leaf rust) strains originating from a random-mating population to determine if unnecessary virulence genes were selected against on susceptible wheat plants. The diverse sexual progeny were cycled for 12 asexual generations through Roblin, Thatcher, and two isogenic Thatcher lines with different known resistance genes. Kolmer found no relationship between the number of unneeded virulence alleles and the fitness of the P. recondita population.

The most susceptible host, Thatcher, maintained the pathogen population with the greatest diversity of virulence alleles, indicating little or no selection against unnecessary virulence alleles. Kolmer suggested that differences in effective population size at the start of the experiment had a significant impact on the results, while selection in favour of necessary specific virulence alleles was more relevant than selection against unnecessary virulence alleles. It is also possible that at least some of these observations were due to selection for particular clones and clonal lineages carrying unnecessary virulence alleles in each host population.

Leonard (1969) used a diverse collection of Puccinia graminis f. sp. avenae (causing oat stem rust) originating from telia-bearing orchard grass stems adjacent to a barberry bush. The diverse pathogen population at the beginning of the experiment was derived from over 300 aecial infections. This rust population was passaged through oat cultivars Craig (susceptible) and Clintland A (with resistance gene A) for eight asexual generations and then tested for presence of virulence alleles corresponding to resistance alleles in a set of oat differentials.

Pustules indicating a resistance reaction (infection types 1 and 2) could be differentiated from pustules indicating a susceptible reaction (infection types 3 and 4). On every differential variety, Leonard found an increase in the frequency of resistant-type pustules with increasing numbers of pathogen generations, consistent with directional selection against pathotypes with unnecessary virulence alleles. He calculated that selection coefficients for unneeded virulence alleles ranged from 0.14 to 0.46 depending on the virulence allele under selection.

Prud'homme and Sackston (1990) used two pathotypes (Race 1 and Race 3) of the sunflower rust pathogen Puccinia helianthi inoculated at equal frequencies onto a susceptible sunflower line (CM 303) and cycled through eight asexual generations. The resulting populations were inoculated onto the susceptible line and two sunflower lines (CM29 and CM90RR) carrying known resistance genes. The simple pathotype (Race 1 with no known virulence alleles) made up 95 per cent of the pathogen population by the end of the 5th generation. In this case, it was not possible to distinguish between selection against unnecessary virulence and selection for a particular genotype.

Bronson and Ellingboe (1986) used two isolates of the wheat powdery mildew pathogen Blumeria graminis to inoculate an isogenic wheat line containing a known resistance gene. The virulence alleles present in the two pathogen strains were assayed on a set of isogenic wheat lines with known resistance genes. Over the course of 12 cycles of asexual reproduction, the frequency of the isolate with more virulence alleles decreased, consistent with a selection coefficient of 0.24 for the isolate with unnecessary virulence alleles.

The two pathogen strains were crossed and the F1 progeny were compared in 8-way mixtures for virulence and reproductive fitness. In the progeny, reduced fitness was found to segregate independently of virulence alleles, suggesting that the fitness of the progeny was not affected by unneeded virulence alleles. The authors concluded that there was no evidence for selection against unneeded virulence and that other genes were more important for determining fitness under these greenhouse conditions.

These examples demonstrate that there are cases where there is and cases where there is no fitness cost associated with maintenance of an avirulence allele in a pathogen population. These examples also highlight how difficult it is to investigate selection and related fitness costs in plant pathogen populations.

Field Experiments to Measure Pathogen Fitness

Field experiments to compare fitness of pathogen strains are much more difficult than greenhouse experiments because it is impossible to prevent immigration of non-inoculants into the experiment and it was difficult to differentiate among strains with certainty before the advent of molecular genetic marker systems. Hence, these experiments remain rare in plant pathology. But these experiments are critical to obtain data on pathogen fitness in actual field environments.

Grant and Archer (1983) used published field survey data of virulence allele frequencies and a simple mathematical model to estimate selection coefficients for particular unneeded virulence alleles in Puccinia graminis f. sp. tritici (causes wheat stem rust) and Blumeria graminis f. sp. hordei (causes barley powdery mildew).

For wheat stem rust, the estimate was based on changes in the frequency of virulence against the Sr6 resistance gene in Australia between 1948 and 1955. For barley powdery mildew, the estimate was based on changes in the frequency of virulence against the Mla6 resistance gene in the United Kingdom between 1969 and 1975. In both plant pathosystems, values for s ranged from 0.04 to 0.06, suggesting that avirulent strains had greater fitness in the absence of corresponding resistance genes.

Using the barley powdery mildew pathosystem (Blumeria graminis f. sp. hordei), Chin and Wolfe (1984) examined the frequencies of pathogen genotypes in naturally infected pure stands and mixtures of four barley cultivars with different resistance genes over a two-year period (1977-78). The virulence complexity of each pathogen population sampled from the field plots was assayed in bulk by inoculating leaf segments of the four component barley cultivars. The authors concluded that in pathogen populations isolated from pure stands, selection favoured strains with fewer virulence alleles, while in host mixtures selection favoured pathogen strains with more virulence alleles. Huang et al. (1994) conducted a similar experiment with this pathosystem in 1989 and 1990 and obtained similar results.

Welz and Leonard (1993) sampled field populations of the maize leaf pathogen Cochliobolus carbonum and used changes in frequencies of three pathotypes over 32 or 51 days to estimate selection coefficients associated with each pathotype. In this pathosystem, pathotypes were differentiated based on lesion morphology and there was not evidence for genetic exchange between pathotypes, though there is some evidence for recombination within pathotypes. Pathotype 3 had lower fitness (s=0.17 on average) than Pathotype 2, and Pathotype 0 had the lowest fitness (s=0.58). The lower fitness of Pathotype 3 was surprising because it produces the largest lesions, but it also produces fewer spores in vitro. This shows that lesion size does not necessarily correlate with overall fitness.

Zhan et al. (2002) used DNA fingerprints to investigate the effect of host resistance and genetic diversity on the population genetic structure of Mycosphaerella graminicola, which causes Septoria tritici leaf blotch on wheat. They collected five strains of the pathogen from each of two wheat cultivars (Stephens and Madsen), and inoculated pure stands and a 1:1 mixture of Madsen and Stephens with an equiproportional mix of spores from all 10 strains at the start of the growing season.

POSTHARVEST PLANT PATHOLOGY

Postharvest diseases affect a wide variety of crops particularly in developing countries which lack sophisticated postharvest storage facilities. Infection by fungi and bacteria may occur during the growing season, at harvest time, during handling, storage, transport and marketing, or even after purchase by the consumer. The reduction of losses in perishable food crops because of postharvest diseases has become a major objective of international organizations. The reality is that there is a portending food crisis that will require the concerted efforts of all who are involved in food production to double their efforts. In fact, to adequately feed the world's expected 10 billion people within the next 40 to 50 years, food production efficiency and distribution will need to be improved immensely.

Specific causes of postharvest losses of fruits and vegetables may be classed as parasitic, non-parasitic, or physical. This chapter deals with the parasitic causes that are of microbiological origin that begin as latent infections before harvest or occur at or after harvest during storage. Fungi are more commonly found attacking fruit and bacteria are more common as postharvest pathogens of vegetables. This chapter will provide a general overview of the subject touching on noteworthy research where it can be used to illustrate postharvest pathology. The reader is encouraged to consult the references for specific information on the topics that are covered.

PREHARVEST FACTORS THAT INFLUENCE POSTHARVEST PATHOLOGY

Postharvest losses vary each year. Prevailing weather while the crop is growing and at harvest contribute greatly to the possibility of decay. Certain cultivars are more prone to decay than are others to specific pathogens. In a recent study, it was found that resistance of major apple cultivars to the fungi that cause blue mold, gray mold, bull's-eye rot, and Mucor rot was dependent on cultivar. Condition of the crop, as determined by fertilizer and soil factors, are very important in susceptibility of the crop to disease. Maturity of the crop at harvest, handling and type of storage have a great deal of influence on how long the crop can be stored without decay.

Weather: Weather affects many factors related to plant diseases, from the amount of inoculum that overwinters successfully to the amount of pesticide residue that remains on the crop at harvest. Abundant inoculum and favourable conditions for infection during the season often result in heavy infection by the time the produce is harvested. For example, conidia of the fungus that causes bull's-eye rot are rain dispersed from cankers and infected bark to fruit especially if rainfall is prolonged near harvest time, causing rotten fruit in cold storage several months later.

Pinpoint or storage scab of apple caused by the same fungus that causes apple scab, and gray mold caused by the fungus Botrytis cinerea, are also very much influenced by the weather. Storage scab only occurs in years with unusually wet Summers and early Falls, when the fruit remain wet for a day or more. These late season infections may not become visible until the apples are in storage. Flowers and fruit are infected by Botrytis cinerea most effectively when it is wet. For example, in grapes infection occurs at 15 to 20°C (59 to 68°F) in the presence of free water after approximately 15 h. In wet seasons strawberries and raspberry crops may be harvested in apparently sound condition, only to decay during tansit and marketing.

Postharvest decay involves further development of pre-harvest infections together with new infections arising from germination of spores on the fruit surface. From these examples it is apparent that decay often has a weather component making thorough weather records an important source of information for predicting possible decay in storage.

Physiological Condition: Condition of produce at harvest determines how long the crop can be safely stored. For example, apples are picked slightly immature to ensure that they can be stored safely for several months. The onset of ripening and senescence in various fruit and vegetables renders them more susceptible to infection by pathogens. On the other hand, fruit and vegetables can be made less prone to decay by management of crop nutrition. For example, calcium has been more closely related to disease resistance than any other cation associated with the cell wall.

In a study on the effect of increased flesh calciumcontent of apples on storage decay fruit were treated with solutions of CaCl2 by dipping, vacuum, or pressure infiltration. Both vacuum and pressure infiltration increased calcium content of the fruit sufficiently to significantly reduce decay. Increased calcium contents in potatoes and peaches have also been documented with reduced postharvest decay. In general, produce containing adequate levels of calcium do not develop physiological disorders and can be stored longer before they breakdown or decay. Conversely, high nitrogen content in fruit predisposes them to decay. In pears, it has been found that management of trees for low nitrogen and high calcium in the fruit reduced severity of postharvest fungal decay. Apple cultivars can be selected for resistance to certain postharvest diseases. For example, 'Royal Gala' is extremely resistant to wound pathogens, 'Granny Smith' to skin punctures, and 'Braeburn' to infiltration of fungal spores into the core.

Fungicide Sprays: Certain pre-harvest sprays are known to reduce decay in storage. Several studies have been done on the effectiveness of pre-harvest ziram fungicide application on pome fruit and show an average reduction in decay of about 25 to 50 per cent with a single spray. Iprodione has been used for several years as a pre-harvest spray 1 day before harvest to prevent infection of stone fruit by Monilinia spp. In combination with wax and/or oil its decay control spectrum is increased and it will also control postharvest fungi such as Rhizopus, and Alternaria. Several new fungicides that are being developed, or have recently been registered promise to protect produce from a number of diseases after harvest. For example, cyprodinil prevented gray mold infection in apple 3 mo after it was applied. The new class of strobilurin fungicides promise to provide postharvest control of several diseases in fruit and vegetables. They are especially effective against fruit scab on apples and should reduce the presence of pin point scab in storage.

Postharvest Factors that Influence Decay

Packing Sanitation: It is important to maintain sanitary conditions in all areas where produce is packed. Organic matter (culls, extraneous plant parts, soil) can act as substrates for decay-causing pathogens. For example, in apple and pear packinghouses, the flumes and dump tank accumulate spores and may act as sources of contamination if steps are not taken to destroy or remove them.

Chlorine readily kills microorganisms suspended in dump tanks and flumes if the amount of available chlorine is adequate. A level of 50 to 100 μL L-1 of active chlorine provides excellent fungicidal activity. Chlorine measured as hypochlorous acid can be obtained by adding chlorine gas, sodium hypochlorite, or dry calcium hypochlorite. Although chlorine effectively kills spores in water it does not protect wounded tissue against subsequent infection from spores lodged in wounds. Organic matter in the water inactivates chlorine, and levels of chlorine must be constantly monitored. The use of a sand filter in association with chlorination improves its efficiency probably because it removes organic matter. Chlorine is sensitive to pH; hypochlorite solutions with higher pH values (7.5 to 8.5) are more stable but less fungicidal, whereas at lower pH values (5.5 to 6.5) the solutions are less stable but more fungicidal.

Recently, chlorine dioxide has replaced chlorine in some sanitizing processes, because several disadvantages limit the use of chlorine, including its unpleasant odour. Chlorine dioxide is not corrosive and is effective over a wide pH range. Recently, in precisely controlled tests in water or as a foam, chlorine dioxide was found to be effective against common postharvest decay fungi on fruit packinghouse surfaces. Peracetic acid is another material that could be used. It has greater stability and faster biocidal properties than chlorine dioxide, but is more corrosive.

The search goes on for effective and economical sanitizing agents. New and old products alike, are continually being evaluated under present day packing operations. Interest in ozone has been rekindled with development of more efficient ozone generators. Acetic acid in the form of a gas is being evaluated for possible use as a sanitizing agent on several crops. It was as effective as SO2 in preventing Gray Mold decay in table grapes stored for 2 mo.

Postharvest Treatments: Products used for postharvest decay control should only be used after the following critical points are considered:

- Type of pathogen involved in the decay.
- Location of the pathogen in the produce.
- Best time for application of the treatment.
- Maturity of the host.

Environment during storage, transportation and marketing of the produce. Specific materials are selected based on these conditions and fall into either chemical or biological categories listed are:

Fungicide treatments: Several fungicides are presently used as postharvest treatments for control of a wide spectrum of decay-causing microorganisms. However, when compared to preharvest pest control products the number is very small. Many former products that were used after harvest are no longer permitted because of concerns with residues and possible toxic effects, the most notable being products that contained benomyl. Other products have been lost as effective controls due to development of resistance by the target pathogen. For example, intensive and continuous use of fungicides for control of blue and green mold on citrus has led to resistance by the causal pathogens of these diseases. Resistance has been reported in many other crops to several different fungicides with different modes of action. Resistance development continues to be an important problem. It has led to the "Fungicide Resistance Action Committee", a cooperative effort between the various producers of fungicides to delay resistance by recommending specific management guidelines.

Examples of postharvest chemical treatments that are presently used are thiabendazole, dichloran, and imazalil. However, resistance to thiabendazole and imazalil is widespread and their use as effective materials is declining. Preservatives or antimicrobial food additives are not generally thought of as postharvest treatments but they do control decay, and in some cases are the only means of control. These products include sodium benzoate, the parabens, sorbic acid, propionic acid, SO2, acetic acid, nitrites and nitrates, and antibiotics such as nisin.

For example, in California gray mold of stored table grapes is prevented by fumigation with SO2. The demand for new postharvest fungicide treatments is strong, especially since the loss of iprodione in 1996. Fludioxinil was granted an emergency registration in 1998 to curb potential losses in nectarines, peaches, and plums that would have resulted. Not all postharvest pathogens are presently controlled by materials that are available. For example, Mucor piriformis, a major postharvest pathogen of apples and Winter pears in the Pacific Northwest is not controlled by any registered fungicide. There is a dire need for new postharvest fungicide treatments that could in part be alleviated by the use of biological control agents.

Biological Control of Postharvest Pathogens: Postharvest biological control is a relatively new approach and offers several advantages over conventional biological control:

- Exact environmental conditions can be established and maintained.
- The biocontrol agent can be targeted much more efficiently.
- Expensive control procedures are cost-effective on harvested food.

Several biological control agents have been developed in recent years, and a few have actually been registered for use on fruit crops. The first biological control agent developed for postharvest use was a strain of Bacillus subtilis. It controlled peach brown rot, but when a commercial formulation of the bacterium was made, adequate disease control was not obtained. More recently, a strain of Pseudomonas syringae van Hall was found that controlled both Blue and Gray Mold of pome fruit. It was subsequently registered, and is now sold commercially for postharvest disease control.

Other bacterial microorganisms are being developed for postharvest disease control. For example, strains of Bacillus pumilus and Pseudomonas fluorescens have been identified that exhibit successful control of B. cinerea in field trials of strawberry. Yeasts such as Pichia guilliermondii and Cryptoccocus laurentii, a yeast that occurs naturally on apple leaves, buds, and fruit were the first to be applied for control of postharvest decay on fruit. The yeast, Candida oleophilia has been registered for control of postharvest decay on fruit crops. The yeasts, Cryptococcus infirmo-minutus and Candida sake successfully control brown rot and blue mold on sweet cherry, and three diseases of apple, respectively, and may be developed commercially.

Although there is no doubt that biocontrols are effective, they do not always give consistent results. This could be because biocontrol efficacy is so directly affected by the amount of pathogen inoculum present. Compatibility with chemicals used during handling is also important. Indications are that biological control agents must be combined with other disease control strategies if they are to provide acceptable control.

Irradiation for Postharvest Decay Ccontrol: Although ultraviolet light has a lethal effect on bacteria and fungi that are exposed to the direct rays, there is no evidence that it reduces decay of packaged fruits and vegetables. More recently, low doses of ultraviolet light (254 nm UV-C) irradiation reduced postharvest brown rot of peaches. In this case, the low dose ultraviolet light treatments had two effects on brown rot development; reduction in the inoculum of the pathogen and induced resistance in the host. However, it has not become a practical postharvest treatment as yet and requires more research. Gamma radiation has been studied for controlling decay, disinfestation, and extending the storage and shelf-life of fresh fruits and vegetables. Dosages of 1.5 to 2 kilogray (kGy), and some cases 3.0 kGy (300 krad), have been effective in controlling decay in several products. A dose of 250 Gy has an adverse effect on grapefruits increasing skin pitting, scald, and decay. Low doses of 150 for fruit flies and 250 gray (Gy) for codling moth are acceptable quarantine procedures. Commercial application of gamma radiation is limited due to the cost and size of equipment needed for the treatment and to uncertainty about the acceptability of irradiated foods to the consumer. Gamma irradiation may be used more in the future once methyl bromide is no longer available to control insect infestation in stored products. All uses of methyl bromide are being phased out to avoid any further damage to the protective layer of ozone surrounding the earth.

Effect of Storage Environment on Postharvest Decay: Commercial producers and handlers modify temperature, RH, and atmospheric composition during prestorage, storage, and transit to control decay. For optimum decay control, two or more factors often are modified simultaneously.

Temperature and RH: Proper management of temperature is so critical to postharvest disease control that all other treatments can be considered as supplements to refrigeration. Fruit rot fungi generally grow optimally at 20 to 25°C (68 to 77°F) and can be conveniently divided into those with a growth minimum of 5 to 10°C (41 to 50°F), or –6 to 0°C (21.2 to 32°F). Fungi with a minimum growth temperature below –2°C (28.4°F) cannot be completely stopped by refrigeration without freezing fruit. However, temperatures as low as possible are desirable because they significantly slow growth and thus reduce decay.

High temperature may be used to control postharvest decay on crops that are injured by low temperatures such as mango, papaya, pepper, and tomato. Although hot water generally is more effective, hot air has been used to control decay in crops that are injured by hot water. Heating of pears at temperatures from 21 to 38°C (69.8 to 100.4°F) for 1 to 7 days reduced postharvest decay. Decay in 'Golden Delicious' apples was reduced by exposure to 38°C (100.4°F) for 4 days and virtually eliminated when treated after inoculation. Heat treatment eliminates incipient infections and improves coverage by fungicides. The primary obstacle to the widespread use of heat to control postharvest fruit diseases or insect infestation is the sensitivity of many fruit to the temperatures required for effective treatment.

Both low and high RH have been related to postharvest decay control. Perforated polyethylene bags for fruit and vegetable storage create RH about 5 to 10 per cent above that in storage rooms. Although shrivel and weight loss are reduced, decay may be increased. Crops such as apples and pears with well-developed cuticle and epidermis, tolerate lower RH levels that help prevent storage decay. Often fungal spore germination is inhibited at low RH, and small differences in RH can have significant effects in relation to the degree of postharvest decay.

Modified or Controlled Atmospheres: Alterations in O_2 and CO_2 concentrations are sometimes provided around fruit and vegetables. With close control of these gases, the synthetic atmosphere is commonly called a controlled atmosphere; the term modified atmosphere is used when there is little possibility of adjusting gas composition during storage or transportation.

Because the pathogen respires as does produce, lowering the O_2 or raising the CO_2 above 5 per cent can suppress pathogenic growth in the host.

In crops such as stone fruits, a direct suppression occurs when fungal respiration and growth are reduced by the high CO_2 of the modified atmosphere. For example, CO_2 added to air has been widely utilized in the transport of 'Bing' cherries, primarily to suppress Gray Mold and Brown Rot.

Low O_2 does not appreciably suppresses fungal growth until the concentration is below 2 per cent. Important growth reductions result if the O_2 is lowered to 1 per cent or lower although there is a danger that the crop will start respiring anaerobically and develop off-flavour. Other technologies that have been tested

for lowering postharvest decay with limited success have been storage and transport under low O_2 and the use of carbon monoxide.

POSTHARVEST DISEASES OF FRUITS

Fruit crops are attacked by a wide range of microorganisms in the postharvest phase. Actual disease only occurs when the attacking pathogen starts to actively grow in the host. Diseases are loosely classified according to their signs and symptoms.

Signs are visible growths of the causal agents, and symptoms the discernible responses produced by the host. In many diseases there is local discolouration and disruption of tissue, with the formation of obvious lesions. Postharvest diseases are caused primarily by microscopic bacteria and fungi, with fungi the most important causal agent in fruit crops. Fungi are further subdivided into classes and are described as lower fungi, characterized by the production of sporangia which give rise to numerous sporangiospores, or higher fungi, described as ascomycetes, deuteromycetes, and basidiomycetes. Ascomycetes are exemplified by fruiting bodies that release sexual spores when mature.

Deuteromycetes, a form of ascomycete, only release asexual spores. They are more common than the sexual ascomycete stage in postharvest crops. Deuteromycetes are further subdivided into hyphomycetes and coelomycetes based on spore and structural characteristics. The agonomycetes contain important soil pathogens that form survival structures known as sclerotia that allow them to survive in the absence of the host. These fungi and the rust and smut fungi are examples of basidiomycetes.

POSTHARVEST DISEASES OF VEGETABLES

Postharvest diseases of vegetables are caused by microscopic fungi and bacteria. Bacteria are more common as pathogens of vegetables than fruit because in general vegetables are less acidic than fruit. They are visible under the light microscope as mostly single-celled rods. Bacteria are capable of very rapid multiplication under the right conditions of pH, temperature, and nutrition. They are classified according to their size, shape, reaction to certain stains, and behaviour on various growth media. The term "vegetable" encompasses a range of plant parts, and the common definition is a culinary one, denoting consumption as a savory rather than as a dessert food. Many vegetables are fruits in the botanical sense, with notable examples being tomatoes, peppers, squashes, and cucumbers.

New Directions for Postharvest Plant Pathology: Postharvest plant pathology has changed its emphasis in recent years. Food safety has emerged as a key element in decay control programmes. Continued failure to effectively control certain postharvest diseases and the need for more environmentally friendly crop control materials is driving a new approach to disease control. Integrated postharvest decay control is the concept that offer the most promise for the future. Society can no longer rely on one or two control strategies but must enlist the entire spectrum of strategies to reduce postharvest losses.

Food Safety Issues: The two most important causes of unsafe food are microbial toxins and contamination of horticultural products by fecal coliforms. The microbial toxins can be subdivided into bacterial toxins and toxins produced by fungi or mycotoxins. An example of a microbial toxin that is extremely toxic are the botulinum toxins produced by the anaerobic bacterium, Clostridium botulinum. Interest in toxins produced by fungi was stimulated by the death of 100,000 turkey poults in England in 1960.

Aflatoxins produced by fungi in the peanut meal used to feed the birds was the cause. Studies have since shown aflatoxins to be potent carcinogens that may occur in nuts and grain. Other toxins have been identified that are produced by the same fungi that cause postharvest decay. For example, patulin produced by Penicillium and Aspergillis spp. can be found in apple and pear products. Patulin is toxic to many biological systems but its role in causing animal and human disease is unclear.

Studies on contamination of horticultural products by fecal coliforms has increased dramatically because of documented incidences of food poisoning from apple juice and seed sprouts. Definite interactions have been shown between plant pathogens and foodborne human pathogens such as Salmonella and Listeria. A study, involving more than 400 samples each of healthy and soft rotted commodities collected in retail markets, indicated that the presence of Salmonella on produce affected by bacterial soft rot was twice that of healthy samples. Controlled experiments with potato, carrot and pepper tissues inoculated with a strain of Salmonella confirmed that bacterial soft rot infection increased multiplication of Salmonella by at least three-to ten-fold compared to multipli-cation on uninfected tissues. Similarly, populations of Listeria monocytogenes, inoculated into decayed apple tissue, continually increased on fruit decayed by Glomerella cingulata, but did not survive after 5 days on fruit decayed by Penicillium expansum.

The pH of the decayed area declined from pH 4.7 to 3.7 in the case of P. expansum, but in the case of G. cingulata pH increased from pH 4.7 to 7.0. This pH modification may be responsible for affecting growth of the foodborne pathogen. Contamination of produce with human pathogens is an important issue that must be addressed while at the same time limiting decay caused by postharvest pathogens and maintaining product quality.

Integrated Control of Postharvest Diseases: Effective and consistent control of storage diseases is dependent upon integration of the following practices:

- Select disease resistant cultivars where possible.
- Maintain correct crop nutrition by use of leaf and soil analysis.
- Irrigate based on crop requirements and avoid overhead irrigation.
- Apply pre-harvest treatments to control insects and diseases.
- Harvest the crop at the correct maturity for storage.
- Apply postharvest treatments to disinfest and control diseases and disorders on produce.
- Maintain good sanitation in packing areas and keep dump water free of contamination.

Store produce under conditions least conducive to growth of pathogens. Integration of cultural methods and biological treatments with yeast biocontrols has been studied on pears. It was found that early harvest, low fruit nitrogen, high fruit calcium, yeast or yeast plus fungicide treatment, and controlled atmosphere storage all reduced severity of blue mold and side rot.

These results demonstrated that unrelated cultural and biological methods that influenced pear decay susceptibility can be combined into an integrated programme to substantially reduce decay.

In another example of an integrated strategy 'Gala' apples were heat treated at 38oC (100.4°F) for 4 days, followed by calcium infiltration with a 2 per cent solution of CaCl2 chloride, and then treated with the microbial antagonist, Pseudomonas syringae. The combined strategy was much more effective than any single strategy for the following reasons. Heat treatment reduced the pathogen population on the fruit surface but did not provide any residual protection. The residual protection was provided by calcium and the biocontrol agent added to the control provided by the heat treatment.

As a general rule, alternatives to chemical control are often less effective than many fungicides. It is highly unlikely that any one alternative method alone will give the same level of control as fungicides. Therefore, it will generally be necessary to combine several alternative methods to develop an integrated strategy to successfully reduce postharvest decay.

GROWTH REGULATORS IN PLANT PATHOLOGY

Symptoms, such as galls, overgrowth or malformation of host tissues, epinasty, elongation, witches broom effects, etc. shows abnormal growth activities in many plant diseases. The extracts of tissues which shows such abnormal growth patterns are often ten times more active in the tests, for growth- regulating substances than are the extracts of healthy plants.

Even in the absence of obvious growth abnormalities; diseased tissues often have increased levels of growth regulators. Four classes of compounds are known to be involved in plant diseases.

All the major types of plants growth regulators (auxins, gibberellins, cytokinins and volatile hormone ethylene) are produced by various viral, bacterial and fungal pathogens. An study of the role of individual growth regulator in pathogenesis is complex because their physiological effects in many cases overlap. Growth regulators in plant diseases have been reviewed by Sequeira.

Auxins

The auxin naturally present in plants is Indole-3-acetic acid (IAA). It is continually produced in growing plant tissues, in which, it is present in concentrations of about 5 μ.g per kg of green plant material. -3gudole Acetic acid) perhaps the nearest to being ubiquitous. Auxin is an organic substance, that promotes growth along the longitudinal axis when applied in low concentrations to the shoots of plants which are freed, as far as possible, from their own inherent growth promoting substances.

"Went", first studied the effects of IAA on Avena, following which it was shown that IAA has a widespread presence in higher plants and especially in diseased tissues. Auxin is also synthesised freely by many species of bacteria and fungi.

The synthesis of IAA in very small amounts takes place in young leaves and root tips. IAA is required for cell elongation and differentiation. Cell elongation is thought to be brought about by the removal of Ca++ and Mg++cation bridges from the pectate network which leads to the formation of straight chain pectic acids or pectinic acids and this in turn allows greater plasticity and extensibility to the cell wall. IAA also facilitates this by forming a complex with, and thus inactivating the enzyme pectin methyl esterase (PME) on the cell membrane and so further stoping the formation of sites suitable for the attachment of divalent cations. The absorption of IAA to the cell membrane also affects the permeability of the membrane. Many basic functions have been attributed to IAA. It has been linked - with cell metabolism, phenol metabolism, increase in respiration, enhancement of the pentose phosphate pathway activity and membrane permeability. Evidence has also accumulated of the primary site of action of IAA being on the nuclear control of RNA synthesis or on RNA itself, thus affecting the level of certain key enzymes in cell growth.

Viral Disease

"Grieve" attributed the reduction in the growth of tomato plants infected with tomato spotted wilt virus to a decrease in auxin content and to decreased rate of translocation through the stem. Surprisingly, Jones did not find a decreased auxin content in the tomato tissues infected with spotted wilt virus. This can be explained by the observation made by Best that such infected plants contain a high level of scopoletin which is a known inhibitor of IAA oxidase. Symptoms of dwarfing in many viral infections are due to the lower- auxin content of the tissues.

Bacterial Diseases

The synthesis and role of auxins in plant diseases has been studied in detail in some bacterial diseases of plants, such as Agrobacterium tumefaciens, Pseudomonas savastoni and Pseudomonas solanacearum. One of the best known plant diseases, displaying overgrowth symptoms, is crown gall caused by Agrobacterium tumefacie.ns. There is enough evidence that the large amounts of auxin found in crown gall tumour are produced by the host.

This disease has been studied by many scientists as a model system in cancer and tumour genesis. Among other things, the pathogen evokes symptoms by altering the hormonal balance of the host cells. Although auxin is produced by A. tumefaciens, the non-auxin producing strains of the organism are also pathogenic and in nature cause galls to be formed.

Thus, the pathogen is able to invade the host to synthasize abnormally high concentrations of auxin. This is accomplished by a pathogen produced factor known as the tumour inducing principle (TIP). While the normal cell is being changed into a tumour cell, its growth increase in an uncontrolled manner and becomes independent of the bacteria. The identity of the transforming agent, TIP, is unknown. TIP has been suggested to be the bacterium, a bacteriophage (a virus associated with the bacterium), a metabolic product of the bacterium, a converted host component, or more recently, bacterial DNA. Although the recent hypothesis that transformation may result from the uptake, incorporation and transformation of a part of the bacterial genome by host cells is attractive, but there is no proof. Equally uncertain is the role of undefined cell division which promotes the factors reported to be present in crown gall tissues.

In main, there is little evidence that the auxins produced by the pathogens play a main part in studying pathogenesis or in determining the nature of plant pathogen interaction. Instead, changes in auxin activity during pathogenesis probably result from alterations in host metabolism, produced by the pathogen.

Fungal Diseases

There are many reports about the production of IAA in bulk from culture media supplemented with tryptophan or other IAA precursors. This has been used as more than circumstantial proof for the production of IAA in vivo by the particular pathogen.

"Shaw" and "Hawkins", found substantial increases (24-times) of auxin corresponding to IAA in Little Club wheat susceptible to and infected with Puccinia graminis tritici. In the case of Atlas barley susceptible to Erysiphe graminis, on the other hand, infected plants showed a maximum five-times increase in IAA in the infected, leaves as compared with the controls. Rust- infected cereals, unlike other rust-diseased hosts, do not show hyperauxiny. enhanced auxin activity has bem demonstrated in the infected tissues in the aecidial stages of Gymnosporangium on Pyrus and Uromyees on Ranunculus.

The auxin content and auxin catabolism of the stems of Eupltorbia cyparissias infected with Uromyces pisi were studied by Pilet. He showed that IAA concentrations in the host are universali associated with the IAA oxidase activity of the host. The inhibition of IAA oxidase may result in the accumulation of auxins in the lesions.

Auxins are also involved in vascular wilt diseases. Beckman showed that cultures of the fungus synthesizes a substance similar to IAA. It was claimed that the synthessizes of this substance in vivo increased the plasticity of xylem parenchyma walls allowing a pectic enzyme weakening of pit membranes and resulting finally in tylosis. IAA solution induced the formation of tylosis in tomato shoots. This supported the view that, auxin is involved in the disease syndrome of the wilt. Increased levels of IAA have been reported in infected stems. Auxins have also been implicated in resistance to ascular wilting. According to Waggoner and Dimond and Corden and Dimond, ethylene, produced in the infected tissues, gives protection from infection.

Vidhyasekaran and Durairaj, discussed the possible reasons for the shot hole symptoms caused in mango leaves, by the pathogen Colletotrichum gloeosporioides. Indole acetic acid and phenols amassed in the lesions, while there was a reduction in IAA content in the tissues around the lesions. Both IAA oxidase and peroxidase activity was inhibited in the lesions while there was an increase in polyphenol oxidase. According to them, tissue breaking is due to the accumulation of auxins in the lesions. The zones around the lesions have less auxin, and hence, uronic carboxyls predominate and the cell walls are rigid. The torn tissue in the lesion per se slough out, leaving the surrounding tissues rigid.

Gibberellins

Gibberellins are the group of chemically similar growth-regulating substances. They can be considered to be normal components of green plants. The existence of gibberellin in plants might still have been unknown if the "bakanae" or "foolish seedling" disease (which had devastating effect on the rice economy of Japan) had not broken out. Japanese farmers noted that plants affected with this disease out-grew normal plants but were thinner and paler than their normal counterparts and there were sometimes lack of fruits. In the early part of this century, Japanese pathologists first showed the connection between the bakanae disease and the fungus called Fusarium moniliforme Sheld.

Sawada proposed that disease may be caused by some product of the fungus. Yabut and Hayshi (1932) were able to seperate the crystalline gibberellin from the infected rice plants. Since that time, gibberellins and alike substances have been found in higher plants and implicated in disease symptoms. All these compounds possess the same general carbon skeleton and all are able to promote either stem elongation or cell division or both in plants. They have a gibbane skeleton.

Numerous functions have been attributed to gibberellins. Thus, gibberellin is the chemical signal, that is, the hormone secreted by the embryo in the seed that catalyses the cells of the aleurone layer into secreting hydrolytic enzymes, such as amylase, which is responsible for liquifying the reserve starch. It also activates the enzymes that promote cell wall break down, thereby aiding in the digestion of the endosperm cells and in weakening the seed-coats which further allows the seeds to germinate.

Besides inducing amylase activity in the aleurone layer of seeds, gibberellins exert a potent stimulus on terminal meristems, and thus in some respects, mimic the effects of auxins on plant growth. Gibberellin also activates or induces the synthesis of other enzymes, including proteinases. Proteinases break down the reserve proteins of the endosperm and release, various amino acids, including tryptophan. Tryptophan is then activated to form auxin (IAA).

Gibberellins, generally have striking growth-promoting effects; they speed the elongation of dwarf varieties to normal size, promote flowering, elongate the stem and root and encourage the growth of fruit. Auxins and gibbe: rellins may also act synergistically. The point of action of the gibberellins in the ceIls'seems to be close to the nucleic acid system, somewhere between DNA and m-RNA. The involvement role of gibberellins in Plan disease has been studied in few diseases. In I. the "bakanae" disease of rice, it is not known whether the infected plants contain more gibberellin than the healthy plants, or in what manner the fungal gibberellin causes the sym. ptoms observed. Following the classical discovery of the "bakanae" disease, the first truely experimental investigations on endogenous gibberellins and infected plants were made by Radley. It was shown by her that yea nodules infected with Rhizobium contain an increased level of GLS as compared to healthy roots.

Gibberellins have been studied more extensively in relation to viral diseases, although no difference has been reported so far in the gibberellin contents of healthy and virus-infected plants. The spraying of diseased plants with gibberellins has been shown to overcome some of the diseases caused by viruses. Thus, the stunting of corn plants infected with corn stunt virus (Maramorosch, 1957) and tobacco plants infected with severe etch virus (Chessin, 1957) was reversed after treatment with gibberellin. Axillary bud suppression caused by sour cherry yellows virus (SCYV) on cherry (Hull and Klos, 1958) and leaf curl virus on tobacco; (Nariani, 1963) was also overcome by gibberellin sprays. The fact that stunting, caused by some viruses, can be reversed by applying exogenous gibberellins suggests that the reduced growth of these virus-infected plant might be due to decreased gibberellin content.

Cytokinins

Cytokinins are the strong growth factors which are must for cell growth and differentiation. The first cytokinin discovered was kinetin by Skoog and Miller in the early 1950s. The main action of cytokinin, is on cell division. It is also capable of orienting plant growth induced by auxin. According to Osborne, cytokinins stimulate protein synthesis at the 1evel of DNA dependent RNA synthesis. Cytokinins also delay death in plant tissues. The action of cytokinins is dependent on auxins.

They are derivatives of adenine. Kinetin (6 furfurylaminopurine), a compound, has been isolated from yeast DNA by Miller. Kinetin was, however, isolated from herring sperm DNA and has not yet been shown to occur naturally in green plants. Several other adenine derivatives with cytokinin like activities have been synthesized and recently, similar compounds, that is, zeatin and N6 (2-iso pentenyl) adenosin (IPA) have been isolated from young corn seeds and from spinach and peas respectively. Cells in cultures show no response when they are supplied with kinetin alone, and they increase in size but do not divide when supplied with only IAA. When both substances arc given together, the cells enlarge and divide.

The relative concentrations of kinetin seem to determine the organs that will differentiate, low concentrations inducing root formation, higher concentrations inducitig bud formation. The mechanism of action of cytokinins is not yet known but the available evidence shows that it is similar to that of gibberellins. Cytokinins act at the stage of m-RNA synthesis, possibly by preventing genes from being turned off and by activating genes that have been previously turned off. The role of cytokinins in plant disease has only now become a subject of study.

A six time increase in cytokinin activity was detected in extracts from the galled roots of root knot nematode-infected tobacco plants compared with extracts from healthy plants; the cytokinin, however, appear late in the infection since they can be detected in galled roots 75 days, and not 35 days, after inoculation. Root cells near a nematode's head display hypertrophy with repeated nuclear divisions together with the incorporation of neighbouring cells, and this results in large, thick-walled multinucleate syncytia or giant cells. The endogenous levels of several hormones are reported to increase in several hosts infected by nematodes. Cytokinin activity also increases in rust- infected beans and broad bean leaves and seems to be related to both the juvenile features of the green islands around the infection centres and the senescenice outside the green island.

In the Helminthosporium blight disease of oats, cytokinins do not enhance the toxicity of toxin victorin but they increase the quantity of toxin absorbed. It appears that cytokinins may increase the effectiveness of the pathogens that produce pathotoxins. On the other hand, cucumber leaf discs treated with kinetin become resistant to the powdery mildew fungus, and tobacco leaves infected with the wild-fire toxin and treated with kinetin fail to develop the typical toxin induced chlorosis. A cytokinin has been implicated in the initiation and development of the fasciation of peas caused by Corynebacterium fasciens. Culture extracts of the pathogens and of infected tissues had materials with cytokinin-like properties. The main active component obtained from the bacterium is, 6-(3 menthyl 2-butanylomino) purine, the riboside of which is found in t-RNA of many bacteria. Plants treated with a high concentration of kinetin attain symptoms similar to those infected by the bacterium. Though these findings suggest that cytokinins play a major role in this disease, a low amount of cytokinin produced by the bacterium in culture indicates that other factors are also involved. Wilson has suggested that cytokinins may be involved in the cell proliferation stage of tumour development in oleander infected with Pseudomonas savas tanoi.

Studies on kinetin and plant virus multiplication have been stimulated by the work of Richmond and Lang, and later by Osborne and Moths concerning the effect of kinetin on protein and nucleic acid metabolism and on amino acids accumulation in leaves. It will be recalled that kinetin tends to stop the breakdown of proteins and nucleic acids which occurs when a leaf is removed from a plant and it tends to mobilize the accumulation of amino acids at the point of its application to a leaf. Osborne has suggested that the new RNA and proteins are even produced under these conditions which, without kinetin, greatly favour catabolism.

There are various conflicting reports of the role of cytokinin in the green island formation in infected leaves. Cornu for the first time studied green island formations in the infections of mildews, downy mildews and rusts. Allen pointed out that the green island formation in mildew infected wheat leaf does not merely represent retention of chlorophyll; they actually are areas in which chlorophyll was destroyed and formed subsequently. The green island formation in Pinto bean leaves infected with Uromyces phaseoli depends on light conditions after the inoculation of the plants. The total green area which finally becomes visible is the result of a arrest in chlorophyll breakdown and the reformation of chlorophyll. Brian speculated that the green island formation may be due to a release of cytokinin by the pathogen. This is, however, doubtful. Most probably the increased cytokinin level starts from host cells and not from the pathogen. Atkin and Neilands (1972) have speculated about the possible role of siderochromes in the formation of green islands.

An increase in cytokinin content has also been noticed in tumours caused by Usti/ago zeae (Turian and Hamilton, 1960) imd Synchytrium endobioticum (Reingard and Pashkar, ~958), and peach leaves infected with Taphrina deformans. However, peach leaf curl symptoms cannot be entirely attributed to changes in cytokinins, as the level of IAA also increases in infected leaves.

Ethylene

Ethylene (C2H4) is a plant growth regulator which is biologically highly active. Ethylene is universally present in plant tissues and is synthesized in increased amounts in damaged or diseased plants. Applied exogenously, it is effective in very low concentrations (below 0.1 ppm) and produces a number of plant responses. Some of these response-epinasty, leaf abscission, premature ageing, fruit ripening and shedding of leaves are similar to the symptoms of certain plant diseases. A number of bacterial or fungal pathogens synthesize ethylene and in the case of fusaria! wilt of tomato caused by F. 'oxysporum f. sp. Iycnpersici, ethylene production by the pathogen is sufficient to account for the epinastic symptoms of a disease. According to Jackson, leaf yellowing and epinasty in waterlogged plants is due to ethylene. Leaf abscission of rose infected with Diplocarpon rosae has been described due to ethylene. This is also connected to auxin action in causing tissue proliferation, increased respiration arid stimulation of root formation. The action of ethylene is due to a change in the permeability of lipoprotein membranes.

Muller demonstrated the production of epinasty by emanations from normal and decayed. Non-virulent isolates of C. fimbriata could induce resistance and it was claimed that more ethylene was synthesized by them than by the virulent isolates.

Sweet potato slices exposed to ethylene (8 ppm for two days) were reported to be rendered resistant to infection and peroxidase and polyphenoloxidase enzyme activity increased during this period. Low concentrations of ethylene were also reported to cause big enhancement in the activity of phenylalanine ammonialyase, a key enzyme in the synthesis of antifungal and antibacterial substances. The enzyme phenylalanine ammonialyase has been implicated in the production of phytoalexins in pea tissue.

According to Daly, both the compatible and incompatible combinations evolve higher amount of ethylene from wheat leaves infected with Puccinia graminis that control leaves, but the most dramatic increase occurs in the compatible reaction at the time when the pathogen starts to sporulate. In the, case of barley leaves infected with Erysiphe graminis a somewhat similar pattern is observed. Peroxidase activity increases in both resistant and susceptible hosts.

Erwinia amylovora does not synthesize ethylene in culture but when cauliflower florets are inoculated with this bacterium, it stimulates ethylene production. This shows that the evolution in diseased tissue most probably is of host origin arising as a result of cellular damage in a still-living tissue. A pectate lyase in the culture filtrate seems to be responsible for the increased production by the plant.

Ethylene undoubtedly interacts with other growth regulating substances, and Van Andel and Fuchs (1972) have pointed out that changes in the balance of growth regulators as a result of such constantly changing interactions may be more important in symptom development than the quantity of any single part. It must be emphasized that it is still not clear whether ethylene is a primary factor in triggering metabolic changes or it is merely produced as a result of these changes.

POPULATION GENETICS AND PHYLOGENETICS IN PLANT PATHOLOGY

Evolutionary biology is a discipline that considers processes that lead to genetic change, or evolution, in populations or species of organisms. Population genetics and phylogenetics are subdisciplines within evolutionary biology. Population genetics considers microevolutionary processes that occur within species and attempts to explain the distribution of genetic variation within and among populations of a single species. Phylogenetics considers macroevolutionary processes that occur among species and attempts to explain the lines of ancestry that led to the current distribution of species across time and space.

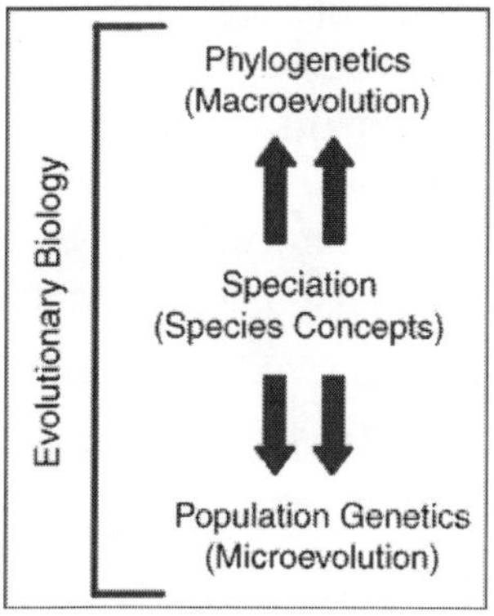

Fig. The Connection between Phylogenetics and Population Genetics Lies in the Process of Speciation.

Evolutionary biology is relevant to plant pathology because pathogen populations evolve in response to control methods and it often is not clear whether a new pathogen population is derived from a pre-existing pathogen population or represents a host-jump or a new species of pathogen. Phylogenetics classifies species based on lines of descent from common ancestors, while population genetics determines relationships among populations within a species. The interface between population genetics and phylogenetics is in the emergence of new species, a process called speciation. Speciation may occur more rapidly for plant pathogens than for other organisms as a result of coevolution.

Phylogenetics and population genetics use different genetic and analytical tools. Phylogenetic analysis is based on using species-specific characters to define taxonomic units based on the methods of cladistics or phenetics. Population genetics is based on using allele frequencies at polymorphic loci to determine genetic structure and define population boundaries. Both areas of research try to elucidate the historical processes and natural history characteristics that led to the current composition of populations and species. Population genetics and phylogenetics often overlap in the fields of biogeography and phylogeography.

When trying to determine the origins of new pathogens, plant pathologists may end up using tools of both phylogenetics and population genetics.

Coevolution Between Plants and Pathogens

When considering plant pathogens, many evolutionary biologists think of plant disease in natural ecosystems as a coevolutionary process. Coevolution is a process during which a trait of one species evolves in response to a trait of another species, which trait in turn evolves in response to the trait in the first species. Coevolution studies consider the reciprocal genetic changes that are expected to occur in two or more ecologically interacting species. The fundamental element of coevolution is reciprocity. A genetic change in one species causes a genetic change in the coevolving species. And this genetic change in turn causes new genetic change in the first species. Reciprocity appears to occur in natural plant pathosystems.

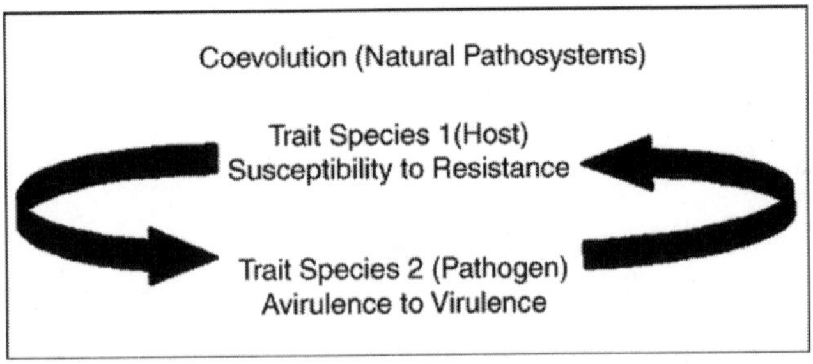

Fig. Coevolution in a Natural Plant Pathosystem is Characterized by the Development of Resistance in the Host and Virulence in the Pathogen.

This coevolution process is thought to occur in natural ecosystems where plant and pathogen exhibit a gene-for-gene (GFG) relationship, as summarized in the quadratic check shown below:

QUADRATIC CHECK IN GFG SYSTEMS

Host		
Pathogen	R_ (resistant)	rr (susceptible)
A (avirulent)	-resistant	+ susceptible
a (virulent)	+ susceptible	+ susceptible

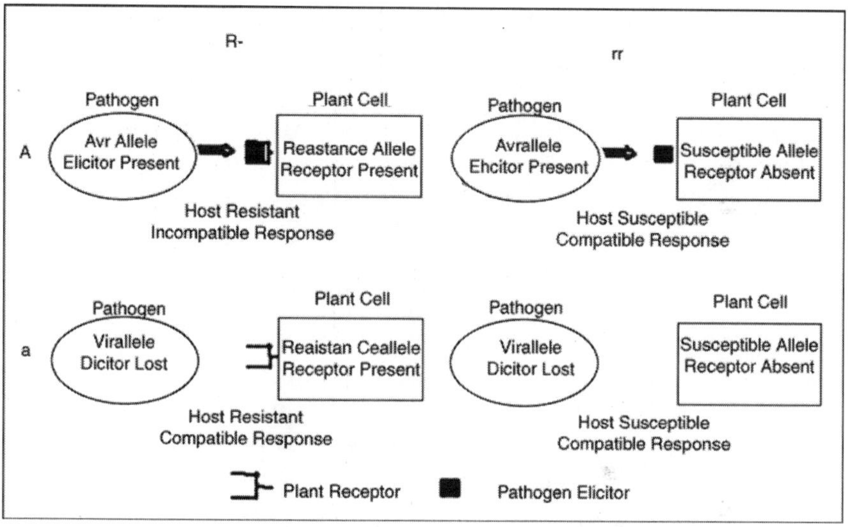

Fig. The Receptor-elicitor Model of Gene-for-gene Interactions. The Resistance Allele of the Plant

Encodes a Receptor that Recognizes an Elicitor Produced by the Pathogen. According to our present knowledge of GFG interactions, the avirulence allele encodes an elicitor that is recognized by a specific receptor (encoded by the R-allele) in the plant host. Recognition of the elicitor by the receptor initiates a signal-transduction pathway that turns on genes involved in the hypersensitive response.

The Logic of GFG Coevolution in Natural Ecosystems

Coevolution begins with a plant population affected by a disease. Within the diseased plant population, a mutation or recombination event occurs that produces a novel receptor allele that recognizes an elicitor in the pathogen. This event causes the mutation from susceptibility to resistance in the plant. In the presence of constant disease pressure that reduces the fitness of susceptible plants, the resistant plants will produce more offspring and thus the resistance allele will increase in frequency.

As the resistance in the plant population increases due to an increase in the frequency of resistant individuals, the ability of the pathogen to reproduce is reduced and the overall level of disease decreases. At this point, there is selection for individuals in the pathogen population that have a mutation causing the loss of the avirulence allele (the elicitor) or a mutation in the avirulence allele that alters the shape of the elicitor so that it is no longer recognized by the plant resistance allele (receptor). This is the mutation from avirulence to virulence in the pathogen.

If the pathogen population has a mutation from avirulence to virulence, there will be selection that favours the virulence allele in the resistant plant population. Once the mutation to virulence occurs, the frequency of the virulence allele increases, leading to an increase in the amount of disease in the previously resistant plant population. And the cycle begins again. Many theories predict that plant alleles for resistance and susceptibility, and pathogen alleles for virulence and avirulence shouldcoexist in natural populations for a long time. Theories also predict that coevolution will occur at the fastest pace in places where the amount of disease is greatest.

Several examples of coevolution exist in natural plant pathosystems. Gene- for-gene interactions between plant resistance genes and pathogen avirulence genes have been demonstrated mainly for obligate biotrophs such as rust fungi and powdery mildew fungi in wild cereal populations. Examples include the wild flax-flax rust fungus populations in Australia (Jeremy J. Burdon and colleagues), and wild oat and oat rust fungus populations in Israel (Isaac Wahl and colleagues).

❐

Chapter - 4

Nature & Importance of Plant Disease

The word 'control' is a complete term where permanent 'control' of a disease is rarely achieved whereas, 'management' of a disease is a continuous process and is more practical in influencing adverse affect caused by a disease. Disease management requires a detail understanding of all aspects of crop production, economics, environmental, cultural, genetics and epidemiological information upon which the management decisions are made.

PRINCIPLES OF PLANT DISEASE MANAGEMENT

There is six basic concept or principles or objectives lying under plant disease management.

- Avoidance of the pathogen: Occurrence of a disease can be avoided by planting/sowing a crop at times when, or in areas where, inoculum remain ineffective/inactive due to environmental conditions, or is rare or absent.
- Exclusion of the pathogen: This can be achieved by preventing the inoculum from entering or establishing in a field or area when it does not exist. Legislative measures like quarantine regulations are needed to be strictly applied to prevent spread of a disease.
- Eradication of the pathogen: It includes reducing, inactivating, eliminating or destroying inoculum at the source, either form a region or from an individual plant (rouging) in which it is already established.
- Protection of the host: Host plants can be protected by creating a toxin barrier on the host surface by the application of chemicals.

- Disease resistance: Preventing infection or reducing the effect of infection of the pathogen through the use of resistance host which is developed by genetic manipulation or by chemotherapy.
- Therapy: Reducing severity of a disease in an infected individual. The first five principles are prophylactic (preventive) procedure and the last one is curative.

Methods of Plant Disease Management

Avoidance of the pathogen:

- Choice of geographical area
- Selection of a field
- Adjustment of time of sowing
- Use of disease escaping varieties
- Use of pathogen-free seed and planting material
- Modification of cultural practices

Exclusion of inoculum of the pathogen

- Treatment of seed and plating materials
- Inspection and certification
- Quarantine regulations
- Eradication of insect vector

Eradication of the pathogen

- Biological control of plant pathogens
- Eradication of alternate and collateral hosts
- Cultural methods:
    - Crop rotation
    - Sanitation of field by destroying/burning crop debris
    - Removal and destruction of diseased plants or plant parts
    - Rouging
- Heat and chemical treatment of diseased plants
- Soil treatment: by use of chemicals, heat energy, flooding and fallowing Protection of the host
- Chemical control: application of chemicals treatment, dusting and spraying
- Chemical control of insect vectors
- Modifications of environment
- Modification of host nutrition

Disease resistance

Use of resistant varieties: Development of resistance in host is done by

- Selection and hybridization for disease resistance.
- Chemotherapy
- Host nutrition
- Genetic Engineering, tissue culture

Therapy

Therapy of diseased plants can be

- Chemotherapy
- Heat therapy
- Tree Surgery

DIAGNOSIS OF PLANT DISEASES

Rapid and accurate diagnosis of disease is necessary before proper control measures can be suggested. It is the first step in the study of any disease. Diagnosis is largely based on characteristic symptoms expressed by the diseased plant. Identification of the pathogen is also essential to diagnosis.

Three steps involved in diagnosis include careful observation and classification of the facts, evaluation of the facts, and a logical decision as to the cause.

Variable Factors Affecting Diagnosis

A skilled diagnostician must know the normal appearance of an affected plant species, its local air and soil environment, the cultural conditions under which it is growing, the pathogens described for the area, and the disease- developing potential of the pathogen. Diagnosis is best done in the presence of the growing plant. Disease is suspected when, for example, part or all of a plant begins to die. Disease also is indicated when blossoms, leaves, stems, roots, or other plant parts appear abnormal—i.e., misshapen, curled, discoloured, overdeveloped, or underdeveloped. Diseased plants also often fail to respond normally to fertilizing, watering, pruning, insect and mite control, or other recommended practices.

Conditions other than infection with a pathogen, however, may produce similar or identical symptoms. Some of these have been described, but numerous other conditions must be considered as well when plants are adversely affected. For example, an affected plant may not be adapted to the area in which it is growing. It may not be able to withstand the extremes in soil moisture, temperature, wind, light, or humidity of the local situation. Damage to plants may be caused by insects, mites, rodents, pets, or humans.

The soil may be poorly drained, gravelly, or overly sandy; it may be covering buried debris—boards, cement blocks, bricks, and mortar; or it may be too dry or otherwise unfavourable for good plant growth. Problems also are caused by high winds, hail, lightning, blowing sand, a heavy load of snow or ice, flooding, fire, ice-removal chemicals, mechanical injury by garden tools or machinery, and fumes from weed-killing chemicals, trash burners, nearby industrial plants, or motor vehicles.

The affected plant may have received treatment different from nearby healthy ones—watering, fertilizing, pest control, pruning, or depth of planting are examples. If different species or kinds of plants in the same area have similar symptoms, the chances are that a pathogen is not involved. Most infectious diseases are highly specific for individual or closely related plant species, and similar symptoms on unrelated plants are usually an indication of some environmental factor rather than a disease-causing organism.

Examination of leaves is usually considered to be the best starting point in diagnosis.

The colour, size, shape, and margins of spots and blights (lesions) are often associated with a particular fungus or bacterium. Many fungi produce "signs" of disease, such as mold growth or fruiting bodies that appear as dark specks in the dead area.

Early stages of bacterial infections that develop on leaves or fruits during humid weather often appear as dark and water-soaked spots with a distinct margin and sometimes a halo—a lighter-coloured ring around the spot.

Low winter temperatures and late spring or early fall freezes cause blasting (sudden death) of leaf and flower buds or sudden blighting (discolouration and death) of tender foliage.

Insect-injured leaves usually show evidence of feeding, such as holes, discolouration, stippling, blotching, downward curling, or other deformations.

Scorching of leaf margins and between the veins is common following hot, dry, windy weather. Similar symptoms are produced by an excess of water, an imbalance of essential nutrients, an excess of soluble salts, changes in the soil water table or soil grade, gas or fume injury, and root injury or disease.

Viral diseases, such as mosaics and yellows, are sometimes confused with injury by a hormone-type weed-killer, unbalanced nutrition, and soil that is excessively alkaline or acid. Nearby plant species are often examined to see if similar symptoms are evident on several different types of plants.

Examination of stems, shoots, branches, and trunk follows a thorough leaf examination. Sunken, swollen, or discoloured areas in the fleshy stem or bark may indicate canker infection by a fungus or bacterium or injury caused by excessively high or low temperatures, hail, tools, equipment, vehicles, or girdling wires.

Fruiting bodies of fungi in or on such areas often indicate secondary infection. Accurate identification of signs as belonging to a pathogenic organism or a secondary or saprophytic one is difficult. Tissues directly infected by pathogenic fungi or bacteria normally show a gradual change in colour or consistency. Injuries, in comparison, are usually well defined with an abrupt change from healthy to affected tissue.

Holes and sawdustlike debris are evidence of boring insects that usually invade woody plants in a low state of vigour. Other borer indications include wilting and dieback (progressive death of shoots that begins at tip and works downward). These symptoms also are produced by fungi and bacteria that invade water- and food-conducting vascular tissue. Symptoms of wilt-inducing microorganisms include dark streaks in sapwood of wilted branches when the wood is cut through at an angle.

Abnormal suckers or water sprouts on trees can indicate careless pruning, extremes in temperature or water supply, structural injury, or disease.

Galls, which are unsightly overgrowths on stem, branch, or trunk, may indicate crown gall, insect injury, water imbalance between plant and soil, or other factors. Crown gall is infectious and develops as rough, roundish galls at wounds, resulting from grafting, pruning, or cultivating.

Wood-decay fungi also enter unprotected wounds, resulting in discoloured, water-soaked, spongy, stringy, crumbly, or hard rots of living and dead wood.

External evidence of wood-decay fungi are clusters of mushrooms (or toadstools) and hoof- or shelf-shaped fungal fruiting structures, called conks, punks, or brackets.

Aboveground symptoms of many root problems look alike. They include stunting of leaf and twig growth, poor foliage colour, gradual or sudden decline in vigour and productivity, shoot wilting and dieback, and even rapid death of the plant. The causes include infectious root and crown rot; nematode, insect, or rodent feeding; low temperature or lightning injury; household gas injury; poor soil type or drainage; change in soil grade; or massive removal of roots in digging utility trenches and construction.

Abnormal root growth is revealed by comparison with healthy roots. Some nematodes, such as root knot (Meloidogyne species), produce small to large galls in roots; other species cause affected roots to become discoloured, stubby, excessively branched, and decayed. Bacterial and fungal root rots commonly follow feeding by nematodes, insects, and rodents.

Diagnosis of a disease complex, one with two or more causes, is usually difficult and requires separation and identification of the individual causes.

Symptoms

The variety of symptoms, the internal and external expressions of disease, that result from any disease form the symptom complex, which, together with the accompanying signs, makes up the syndrome of the disease.

Generalized symptoms may be classified as local or systemic, primary or secondary, and microscopic or macroscopic. Local symptoms are physiological or structural changes within a limited area of host tissue, such as leaf spots, galls, and cankers.

Systemic symptoms are those involving the reaction of a greater part or all of the plant, such as wilting, yellowing, and dwarfing. Primary symptoms are the direct result of pathogen activity on invaded tissues (e.g., swollen "clubs" in clubroot of cabbage and "galls" formed by feeding of the root-knot nematode).

Secondary symptoms result from the physiological effects of disease on distant tissues and uninvaded organs (e.g., wilting and drooping of cabbage leaves in hot weather resulting from clubroot or root knot). Microscopic disease symptoms are expressions of disease in cell structure or cell arrangement seen under a microscope.

Macroscopic symptoms are expressions of disease that can be seen with the unaided eye. Specific macroscopic symptoms are classified under one of four major categories: prenecrotic, necrotic, hypoplastic, and hyperplastic or hypertrophic. These categories reflect abnormal effects on host cells, tissues, and organs that can be seen without a hand lens or microscope.

NATURE AND IMPORTANCE OF PLANT DISEASES

Plant diseases are known from times preceding the earliest writings. Fossil evidence indicates that plants were affected by disease 250 million years ago. The Bible and other early writings mention diseases, such as rusts, mildews, blights, and blast, that have caused famine and other drastic changes in the economy of nations since the dawn of recorded history. Other plant disease outbreaks with similar far-reaching effects in more recent times include late blight of potato in Ireland; powdery and downy mildews of grape in France; coffee rust in Ceylon; Fusarium wilts of cotton and flax; southern bacterial wilt of tobacco (early 1900s); Sigatoka leaf spot and Panama disease of banana in Central America; black stem rust of wheat; and southern corn leaf blight in the United States.

Loss of crops from plant diseases may result in hunger and starvation, especially in less developed countries where access to disease-control methods is limited and annual losses of 30 to 50 per cent are common for major crops. In some years, losses are much greater, producing catastrophic results for those who depend on the crop for food. Major disease outbreaks among food crops have led to famines and mass migrations throughout history. The devastating outbreak of late blight of potato (Phytophthora infestans) that began in Europe in 1845 and brought about the Irish famine caused starvation, death, and mass migration of the Irish population.

Of a population of eight million, approximately one million (about 12.5 per cent) died of starvation and 1.5 million (almost 19 per cent) emigrated, mostly to the United States, as refugees from the destructive blight. This fungus thus had a tremendous influence on the economic, political, and cultural development in Europe and the United States. During World War I, late blight damage to the potato crop in Germany may have helped end the war. Losses from plant diseases also can have a significant economic impact, causing a reduction in income for crop producers and distributors and higher prices for consumers. In 1993 the United States lost more than one million acres (405,000 hectares) of crops to disease. More than 800,000 acres of wheat succumbed to disease, exacting a monetary loss in the millions of dollars.

Plant diseases are a normal part of nature and one of many ecological factors that help keep the hundreds of thousands of living plants and animals in balance with one another. Plant cells contain special signaling pathways that enhance their defences against insects, animals, and pathogens. One such example involves a plant hormone called jasmonate (jasmonic acid). In the absence of harmful stimuli, jasmonate binds to special proteins, called JAZ proteins, to regulate plant growth, pollen production, and other processes. In the presence of harmful stimuli, however, jasmonate switches its signaling pathways, shifting instead to directing processes involved in boosting plant defence. Genes that produce jasmonate and JAZ proteins represent potential targets for genetic engineering to produce plant varieties with increased resistance to disease.

Humans have carefully selected and cultivated plants for food, clothing, shelter, fibre, and beauty for thousands of years. Disease is just one of many hazards that must be considered when plants are taken out of their natural environment and grown in pure stands under what are often abnormal conditions.

Many valuable crop and ornamental plants are very susceptible to disease and would have difficulty surviving in nature without human intervention. Cultivated plants are often more susceptible to disease than are their wild relatives. This is because large numbers of the same species or variety, having a uniform genetic background, are grown close together, sometimes over many thousands of square kilometres. A pathogen may spread rapidly under these conditions.

Definitions of Plant Disease

In general, a plant becomes diseased when it is continuously disturbed by some causal agent that results in an abnormal physiological process that disrupts the plant's normal structure, growth, function, or other activities. This interference with one or more of a plant's essential physiological or biochemical systems elicits characteristic pathological conditions or symptoms.

Plant diseases can be broadly classified according to the nature of their primary causal agent, either infectious or non-infectious. Infectious plant diseases are caused by a pathogenic organism such as a fungus, bacterium, mycoplasma, virus, viroid, nematode, or parasitic flowering plant. An infectious agent is capable of reproducing within or on its host and spreading from one susceptible host to another.

Non-infectious plant diseases are caused by unfavourable growing conditions, including extremes of temperature, disadvantageous relationships between moisture and oxygen, toxic substances in the soil or atmosphere, and an excess or deficiency of an essential mineral. Because non-infectious causal agents are not organisms capable of reproducing within a host, they are not transmissible.

In nature, plants may be affected by more than one disease-causing agent at a time. A plant that must contend with a nutrient deficiency or an imbalance between soil moisture and oxygen is often more susceptible to infection by a pathogen; a plant infected by one pathogen is often prone to invasion by secondary pathogens. The combination of all disease-causing agents that affect a plant make up the disease complex. Knowledge of normal growth habits, varietal characteristics, and normal variability of plants within a species—as these relate to the conditions under which the plants are growing—is required for a disease to be recognized.

The study of plant diseases is called plant pathology. Pathology is derived from the two Greek words pathos (suffering, disease) and logos (discourse, study). Plant pathology thus means a study of plant diseases.

DISEASE DEVELOPMENT AND TRANSMISSION

Pathogenesis and Saprogenesis

Pathogenesis is the stage of disease in which the pathogen is in intimate association with living host tissue. Three fairly distinct stages are involved:

Inoculation: Transfer of the pathogen to the infection court, or area in which invasion of the plant occurs (the infection court may be the unbroken plant surface, a variety of wounds, or natural openings—e.g., stomates [microscopic pores in leaf surfaces], hydathodes [stomatelike openings that secrete water], or lenticels [small openings in tree bark])

Incubation: The period of time between the arrival of the pathogen in the infection court and the appearance of symptoms.

Infection: The appearance of disease symptoms accompanied by the establishment and spread of the pathogen.

One of the important characteristics of pathogenic organisms, in terms of their ability to infect, is virulence. Many different properties of a pathogen contribute to its ability to spread through and to destroy the tissue. Among these virulence factors are toxins that kill cells, enzymes that destroy cell walls, extracellular polysaccharides that block the passage of fluid through the plant system, and substances that interfere with normal cell growth.

Not all pathogenic species are equal in virulence—that is, they do not produce the same amounts of the substances that contribute to the invasion and destruction of plant tissue. Also, not all virulence factors are operative in a particular disease. For example, toxins that kill cells are important in necrotic diseases, and enzymes that destroy cell walls play a significant role in soft rot diseases.

Many pathogens, especially among the bacteria and fungi, spend part of their life cycles as pathogens and the remainder as saprophytes.

Saprogenesis is the part of the pathogen's life cycle when it is not in vital association with living host tissue and either continues to grow in dead host tissue or becomes dormant. During this stage, some fungi produce their sexual fruiting bodies; the apple scab (Venturia inaequalis), for example, produces perithecia, flask-shaped spore-producing structures, in fallen apple leaves.

Other fungi produce compact resting bodies, such as the sclerotia formed by certain root- and stem-rotting fungi (Rhizoctonia solani and Sclerotinia sclerotiorum) or the ergot fungus (Claviceps purpurea). These resting bodies, which are resistant to extremes in temperature and moisture, enable the pathogen to survive for months or years in soil and plant debris in the absence of a living host.

Epiphytotics

When the number of individuals a disease affects increases dramatically, it is said to have become epidemic (meaning "on or among people"). A more precise term when speaking of plants, however, is epiphytotic ("on plants"); for animals, the corresponding term is epizootic. In contrast, endemic (enphytotic) diseases occur at relatively constant levels in the same area each year and generally cause little concern.

Epiphytotics affect a high percentage of the host plant population, sometimes across a wide area. They may be mild or destructive and local or regional in occurrence. Epiphytotics result from various combinations of factors, including the right combination of climatic conditions. An epiphytotic may occur when a pathogen is introduced into an area in which it had not previously existed. Examples of this condition include the downy mildews (Sclerospora species) and rusts (Puccinia species) of corn in Africa during the 1950s, the introduction of the coffee rust fungus into Brazil in the 1960s, and the entrance of the chestnut blight (Endothia parasitica) into the United States shortly after 1900.

Also, when new plant varieties are produced by plant breeders without regard for all enphytotic diseases that occur in the same area to some extent each year (but which are normally of minor importance), some of these varieties may prove very susceptible to previously unimportant pathogens. Examples of this situation include the development of oat varieties with Victoria parentage, which, although highly resistant to rusts (Puccinia graminis avenae and P. coronata avenae) and smuts (Ustilago avenae, U. kolleri), proved very susceptible to Helminthosporium blight (H. victoriae), formerly a minor disease of grasses.

The destructiveness of this disease resulted in a major shift of oat varieties on 50 million acres in the United States in the mid-1940s. Corn (maize) with male-sterile cytoplasm (i.e., plants with tassels that do not extrude anthers or pollen), grown on 60 million acres in the United States, was attacked in 1970 by a virulent new race of the southern corn leaf blight fungus (Helminthosporium maydis race T), resulting in a loss of about 700 million bushels of corn. More recently the new Helminthosporium race was widely disseminated and was reported from most continents. Finally, epiphytotics may occur when host plants are cultivated in large acreages where previously little or no land was devoted to that crop.

Epiphytotics may occur in cycles. When a plant disease first appears in a new area, it may grow rapidly to epiphytotic proportions. In time, the disease wanes, and, unless the host species has been completely wiped out, the disease subsides to a low level of incidence and becomes enphytotic. This balance may change dramatically by conditions that favour a renewed epiphytotic. Among such conditions are weather (primarily temperature and moisture), which may be very favourable for multiplication, spread, and infection by the pathogen; introduction of a new and more susceptible host; development of a very aggressive race of the pathogen; and changes in cultural practices that create a more favourable environment for the pathogen.

Environmental Factors Affecting Disease Development

Important environmental factors that may affect development of plant diseases and determine whether they become epiphytotic include temperature, relative humidity, soil moisture, soil pH, soil type, and soil fertility.

Temperature

Each pathogen has an optimum temperature for growth. In addition, different growth stages of the fungus, such as the production of spores (reproductive units), their germination, and the growth of the mycelium (the filamentous main fungus body), may have slightly different optimum temperatures. Storage temperatures for certain fruits, vegetables, and nursery stock are manipulated to control fungi and bacteria that cause storage decay, provided the temperature does not change the quality of the products. Little, except limited frost protection, can be done to control air temperature in fields, but greenhouse temperatures can be regulated to check disease development.

Knowledge of optimum temperatures, usually combined with optimum moisture conditions, permits forecasting, with a high degree of accuracy, the development of such diseases as blue mold of tobacco (Peronospora tabacina), downy mildews of vine crops (Pseudoperonospora cubensis) and lima beans (Phytophthora phaseoli), late blight of potato and tomato (Phytophthora infestans), leaf spot of sugar beets (Cercospora beticola), and leaf rust of wheat (Puccinia recondita tritici). Effects of temperature may mask symptoms of certain viral and mycoplasmal diseases, however, making them more difficult to detect.

Relative humidity

Relative humidity is very critical in fungal spore germination and the development of storage rots. Rhizopus soft rot of sweet potato (Rhizopus stolonifer) is an example of a storage disease that does not develop if relative humidity is maintained at 85 to 90 per cent, even if the storage temperature is optimum for growth of the pathogen. Under these conditions, the sweet potato root produces suberized (corky) tissues that wall off the Rhizopus fungus.

High humidity favours development of the great majority of leaf and fruit diseases caused by fungi and bacteria. Moisture is generally needed for fungal spore germination, the multiplication and penetration of bacteria, and the initiation of infection. Germination of powdery mildew spores occurs best at 90 to 95 per cent relative humidity. Diseases in greenhouse crops—such as leaf mold of tomato (Cladosporium fulvum) and decay of flowers, leaves, stems, and seedlings of flowering plants, caused by Botrytis species—are controlled by lowering air humidity or by avoiding spraying plants with water.

Soil Moisture

High or low soil moisture may be a limiting factor in the development of certain root rot diseases. High soil-moisture levels favour development of destructive water mold fungi, such as species of Aphanomyces, Pythium, and Phytophthora. Excessive watering of houseplants is a common problem. Overwatering, by decreasing oxygen and raising carbon dioxide levels in the soil, makes roots more susceptible to root-rotting organisms.

Diseases such as take-all of cereals (Ophiobolus graminis); charcoal rot of corn, sorghum, and soybean (Macrophomina phaseoli); common scab of potato (Streptomyces scabies); and onion white rot (Sclerotium cepivorum) are most severe under low soil-moisture levels.

Soil pH

Soil pH, a measure of acidity or alkalinity, markedly influences a few diseases, such as common scab of potato and clubroot of crucifers (Plasmodiophora brassicae). Growth of the potato scab organism is suppressed at a pH of 5.2 or slightly below (pH 7 is neutral; numbers below 7 indicate acidity, and those above 7 indicate alkalinity). Scab is not normally a problem when the natural soil pH is about 5.2. Some farmers add sulfur to their potato soil to keep the pH about 5.0. Clubroot of crucifers (members of the mustard family, including cabbage, cauliflower, and turnips), on the other hand, can usually be controlled by thoroughly mixing lime into the soil until the pH becomes 7.2 or higher.

Soil Type

Certain pathogens are favoured by loam soils and others by clay soils. Phymatotrichum root rot attacks cotton and some 2,000 other plants in the southwestern United States. This fungus is serious only in black alkaline soils— pH 7.3 or above—that are low in organic matter. Fusarium wilt disease, which attacks a wide range of cultivated plants, causes more damage in lighter and higher (topographically) soils. Nematodes are also most damaging in lighter soils that warm up quickly.

Soil Fertility

Greenhouse and field experiments have shown that raising or lowering the levels of certain nutrient elements required by plants frequently influences the development of some infectious diseases—for example, fire blight of apple and pear, stalk rots of corn and sorghum, Botrytis blights, Septoria diseases, powdery mildew of wheat, and northern leaf blight of corn. These diseases and many others are more destructive after application of excessive amounts of nitrogen fertilizer. This condition can often be counteracted by adding adequate amounts of potash, a fertilizer containing potassium.

Requirements for Disease Development

Infectious disease cannot develop if any one of the following three basic conditions is lacking: (1) the proper environment, the most important environmental factors being the amount and frequency of rains or heavy dews, the relative humidity, and the air and soil temperatures, (2) the presence of a virulent pathogen, and (3) a susceptible host. Effective disease-control measures are aimed at breaking this environment-pathogen-host triangle. Loss resulting from disease is reduced, for example, if the host can be made more resistant or immune through such techniques as plant breeding or genetic engineering.

In addition, the environment can be made less favourable for invasion by the pathogen and more favourable for the growth of the host plant. Finally, the pathogen can be killed or prevented from reaching the host. These basic methods of control can be divided into a number of cultural, chemical, and biological practices to help control the disease.

CLASSIFICATION OF PLANT DISEASES BY CAUSAL AGENT

Plant diseases are often classified by their physiological effects or symptoms. Many diseases, however, produce practically identical symptoms and signs but are caused by very different microorganisms or agents, thus requiring completely different control methods. Classification according to symptoms is also inadequate because a causal agent may induce several different symptoms, even on the same plant organ, which often intergrade. Classification may be according to the species of plant affected. Host indexes (lists of diseases known to occur on certain hosts in regions, countries, or continents) are valuable in diagnosis. When an apparently new disease is found on a known host, a check into the index for the specific host often leads to identification of the causal agent. It is also possible to classify diseases according to the essential process or function that is adversely affected. The best and most widely used classification of plant diseases is based on the causal agent, such as a non-infectious agent or an infectious agent (i.e., a virus, viroid, mycoplasma, bacterium, fungus, nematode, or parasitic flowering plant).

Non-infectious Disease-causing Agents

Non-infectious diseases, which sometimes arise very suddenly, are caused by the excess, deficiency, non-availability, or improper balance of light, air circulation, relative humidity, water, or essential soil elements; unfavourable soil moisture-oxygen relations; extremes in soil acidity or alkalinity; high or low temperatures; pesticide injury; other poisonous chemicals in air or soil; changes in soil grade; girdling of roots; mechanical and electrical agents; and soil compaction. In addition, unfavourable preharvest and storage conditions for fruits, vegetables, and nursery stock often result in losses. The effects of non-infectious diseases can be seen on a variety of plant species growing in a given locality or environment. Many diseases and injuries caused by non- infectious agents result in heavy loss but are difficult to check or eliminate because they frequently reflect ecological factors beyond human control. Symptoms may appear several weeks or months after an environmental disturbance.

Injuries incurred from accidents, poisons, or adverse environmental disturbances often result in damaged tissues that weaken a plant, enabling bacteria, fungi, or viruses to enter and add further damage. The cause may be obvious (lightning or hail), but often it is obscure. Symptoms alone are often unreliable in identifying the causal factor. A thorough examination of recent weather patterns, the condition of surrounding plants, cultural treatments or disturbances, and soil and water tests can help reveal the nature of the disease.

Adverse Environment

High temperatures may scald corn, cotton, and bean leaves and may induce formation of cankers at the soil surface of tender flax, cotton, and peanut plants. Frost injury is relatively common, but temperatures just above freezing also may cause damage, such as net necrosis (localized tissue death) in potato tubers and "silvering" of corn leaves. Isolated, thin-barked trees growing in northern climates and subjected to frequent thawing by day and freezing by night may develop dead bark cankers or vertical frost cracks on the south or southwest sides of the trunk. Alternate freezing and thawing, heaving, low air moisture, and smothering under an ice-sheet cover are damaging to alfalfa, clovers, strawberries, and grass on golf greens. Legume crowns commonly split under these conditions and are invaded by decay-forming fungi.

The drought and dry winds that often accompany high temperatures cause stunting, wilting, blasting, marginal scorching of leaves, and dieback of shoots. Leaf scorch is common on trees in exposed locations following hot, dry, windy weather when water is lost from leaves faster than it is absorbed by roots. Leaf scorch and sudden flower drop are common indoor plant problems because the humidity in a home, an apartment, or an office is usually below 30 per cent. Similar symptoms are caused by a change in soil grade, an altered water-table level, a compacted and shallow soil, paved surface over tree roots, temporary flooding or a waterlogged (oxygen-deficient) soil, girdling tree roots, salt spray near the ocean, and an injured or diseased root system. Injured plants are often very susceptible to air and soil pathogens and secondary invaders.

Blossom-end rot of tomato and pepper is prevalent when soil moisture and temperature levels fluctuate widely and calcium is low.Poor aeration may cause blackheart in stored potatoes. Accumulation of certain gases from the respiration of apples in storage may produce apple scald and other disorders.

All plants require certain mineral elements to develop and mature in a healthy state. Macronutrients such as nitrogen, potassium, phosphorus, sulfur, calcium, and magnesium are required in substantial quantities, while micronutrients or trace elements such as boron, iron, manganese, copper, zinc, and molybdenum are needed in much smaller quantities. When the supply of any essential nutrient falls below the level required by the plant, a deficiency occurs, leading to symptoms that include stunting of plants; scorching or malformation of leaves; abnormal colouration; premature leaf, bud, and flower drop; delayed maturity or failure of flower and fruit buds to develop; and dieback of shoots.

Symptoms of nutrient deficiencies vary depending on the nutrients involved, the stage of plant growth, soil moisture, and other factors; they often resemble symptoms caused by infectious agents such as bacteria or viruses.

The availability of water may affect nutrient uptake by the plant. Blossom- end rot of tomato, a disease associated with a deficiency of calcium, may occur if the water supply is irregular, even if an adequate amount of calcium is in the soil. This discontinuity in availability of water will inhibit uptake of the calcium in a quantity sufficient to nourish a fast-growing tomato plant. Necrosis at the blossom end of the fruit results. This situation generally disappears when water conditions improve. Excess minerals can damage plants either directly, causing stunting, deformities, or dieback, or indirectly by interfering with the absorption and use of other nutrients, resulting in subsequent deficiency symptoms. A superabundance of nitrogen, for example, may cause deficiency symptoms of potassium, zinc, or other nutrient elements; a lack of or delay in flower and fruit development; and a predisposition to winter injury. If potassium is high, calcium and magnesium deficiencies may occur.

The pH of a soil has a dramatic impact on nutrient availability to plants. Most plants will grow in a soil with a pH between 4.0 and 8.0. In acidic soils some nutrients are far more available and may reach concentrations that are toxic or that inhibit absorption of other nutrients, while other minerals become chemically bound and unavailable to plants. A similar situation exists in alkaline soils, although different minerals are affected. Oats planted in alkaline soils that actually contain a sufficient amount of manganese may develop the manganese-deficiency disease gray speck. This occurs because an elevated soil pH causes manganese to react with oxygen to produce manganese dioxide, a form of the nutrient that is insoluble to plants.

An excess of water-soluble salts is a common problem with houseplants. Salt concentrations may build up as a whitish crust on soil and container surfaces of potted plants following normal evaporation of water over a period of time. Symptoms include leaf scorching, bronzing, yellowing and stunting, and wilting, plus root and shoot dieback. Damage from soluble salts is also common in arid regions and in regions where ice-control chemicals are applied heavily.

Several non-parasitic diseases (e.g., oat blast, weakneck of sorghum, straighthead of rice, and crazy-top of cotton) are caused by combinations of environmental factors—e.g., high temperatures, moisture stress or poor irrigation practices, imbalance of mineral nutrients, and reduced light.

Environmental disturbances alter the normal physiology of the plant, activity of pathogens, and host-pathogen interactions.

Toxic Chemicals

Many complex chemicals are routinely applied to plants to prevent attack by insects, mites, and pathogens; to kill weeds; or to control growth. Serious damage may result when fertilizers, herbicides, fumigants, growth regulators, antidesiccants, insecticides, miticides, fungicides, nematicides, and surfactants (substances with enhanced wetting, dispersing, or cleansing properties, such as detergents) are applied at excessive rates or under hot, cold, or slow-drying conditions.

Some pollutants are the direct products of industry and fuel combustion, while others are the result of photochemical reactions between products of combustion and naturally occurring atmospheric compounds. The major pollutants toxic to plants are sulfur dioxide, fluorine, ozone, and peroxyacetyl nitrate. Sulfur dioxide results primarily from the burning of large amounts of soft coal and high-sulfur oil. It is toxic to a wide range of plants at concentrations as low as 0.25 part per million (ppm) of air (i.e., on a volume basis, one part per million represents one volume of pure gaseous toxic substance mixed in one million volumes of air) for 8 to 24 hours. Gaseous and particulate fluorides are more toxic to sensitive plants than is sulfur dioxide because they are accumulated by leaves. They are also toxic to animals that feed on such foliage. Fluorine injury is common near metal-ore smelters, refineries, and industries making fertilizers, ceramics, aluminum, glass, and bricks.

Ozone and peroxyacetyl nitrate injury (also called oxidant injury) are more prevalent in and near cities with heavy traffic problems. Exhaust gases from internal combustion engines contain large amounts of hydrocarbons (substances that principally contain carbon and hydrogen molecules—

gasoline, for example). Smaller amounts of unconsumed hydrocarbons are formed by combustion of fossil fuels (e.g., coal, oil, natural gas) and refuse burning. Ozone, peroxyacetyl nitrate, and other oxidizing chemicals (smog) are formed when sunlight reacts with nitrogen oxides and hydrocarbons. This pollutant complex is damaging to susceptible plants many kilometres from its source. Ozone and peroxyacetyl nitrate are capable of causing injury if present at levels of 0.01 to 0.05 part per million for several hours.

Physical Injury

Lightning, hail, high winds, ice and snow loads, machinery, insect and animal feeding, and various cultural practices may seriously injure plants or plant products. With the exception of lightning, which may cause death of trees and succulent crop plants in limited areas, such injury does not usually kill plants. Wounds are created, however, through which pathogens may enter.

Infectious Disease-causing Agents

Plants are subject to infection by thousands of species from very diverse groups of organisms. Most are microscopic, but a few are macroscopic. The infectious agents, as previously mentioned, are called pathogens and can be grouped as follows: viruses and viroids, bacteria (including mycoplasmas and spiroplasmas, collectively referred to as mycoplasma-like organisms [MLOs]), fungi, nematodes, and parasitic seed plants.

Diseases Caused by Viruses and Viroids

General Characteristics

Viruses and viroids are the smallest of the infectious agents. The structurally mature infectious particle is called a virion. Virions range in size from approximately 20 nanometres (0.0000008 inch) to 250–400 nanometres and are of various shapes. Viroids differ from viruses in that they have no structural proteins, such as those that form the protein coat (capsid) of the virus. Both viruses and viroids are obligate parasites—i.e., they are able to multiply or replicate only within a living cell of a particular host. A single plant species may be susceptible to infection by several different viruses or viroids. Major disease of important food crops such as potato, tomato, wheat, oats, rice, corn, peach, orange, sugar beet, sugarcane, and palm result from viral infection. Diseases are generally most serious in plants that are propagated vegetatively, or asexually—i.e., grown from cuttings, cut divisions, sprouts, and other plant material—rather than grown from seeds (sexually propagated).

Symptoms

The symptoms of viral and viroid plant diseases fall into four groups: (1) change in colour—yellowing, green and yellow mottling, and vein clearing; (2) malformations—distortion of leaves and flowers, rosetting, proliferation and witches'-brooms (abnormal proliferation of shoots), and little or no leaf development between the veins; (3) necrosis—leaf spots, ring spots, streaks, wilting or drooping, and internal death, especially of phloem (food-conducting) tissue; and (4) stunting or dwarfing of leaves, stems, or entire plants. Rarely they may kill the host in a short time (e.g., spotted wilt and curly top of tomato). More commonly they cause reduced yield and lower quality of product.

In many cases, virus-infected plants are more susceptible to root rots, stem or stalk rots, seedling blights, and possibly other types of diseases.

Some plants may carry one or more viruses and show no symptoms; thus, they are latent carriers and a source of infection for other plants. Symptoms of certain virus-infected plants, such as geraniums, may be masked at high temperatures. Virus symptoms reappear when the weather cools.

For convenience, viral/viroid diseases are often grouped together generally by symptoms, regardless of true viral/viroid relationships. Viruses also can be grouped into strains, each differing greatly in virulence and other properties. For example, two virus strains, chemically distinct, may produce indistinguishable symptoms in one orchid plant but strikingly different symptoms in another. Diseases caused by unrelated viruses may resemble one another more closely than diseases caused by strains of the same virus. Certain variegated plants, such as Abutilon and Rembrandt tulips, owe their horticultural uniqueness and desirability to being inherently virus-infected.

Transmission

With the exception of tobacco mosaic virus, relatively few viruses or viroids are spread extensively in the field by contact between diseased and healthy leaves. All viruses that spread within their host tissues (systemically) can be transmitted by grafting branches or buds from diseased plants on healthy plants. Natural grafting and transmission are possible by root grafts and with dodder (Cuscuta species). Vegetative propagation often spreads plant viruses. Fifty to 60 viruses are transmitted in seed, and a few seed-borne viruses, such as sour-cherry yellows, are carried in pollen and transmitted by insects.

Most disease-causing viruses are carried and transmitted naturally by insects and mites, which are called vectors of the virus. The principal virus- carrying insects are about 200 species of aphids, which transmit mostly mosaic viruses, and more than 100 species of leafhoppers, which carry yellows-type viruses. Whiteflies, thrips, mealybugs, plant hoppers, grasshoppers, scales, and a few beetles also serve as vectors for certain viruses. Some viruses may persist for weeks or months and even duplicate themselves in their insect vectors; others are carried for less than an hour. Slugs, snails, birds, rabbits, and dogs also transmit a few viruses, but this is not common.

A small number of plant viruses are soilborne. Viruses causing grape fanleaf, tobacco rattle, and tobacco and tomato ring spots, as well as several strawberry viruses, are spread by nematodes feeding externally (i.e., ectoparasitic) on plant roots. A few soilborne viruses may be spread by the swimming spores of primitive, soil-inhabiting pathogenic fungi, such as those causing big vein of lettuce, soilborne wheat mosaic, and tobacco necrosis. Viruses often overwinter in biennial and perennial crops and weeds (plants that overwinter by means of roots and produce seed in their second year or during several years, respectively), in plant debris, and in insect vectors. Plants, once infected, normally remain so for life.

Control

After a plant is infected with a virus/viroid, little can be done to restore its health. Control is accomplished by several methods, such as growing resistant species and varieties of plants or obtaining virus-free seed, cuttings, or plants as a result of indexing and certification programmes. Indexing is a procedure to determine the presence or absence of viruses not readily transmitted mechanically. Material from a "test" plant is grafted to an "indicator" plant that develops characteristic symptoms if affected by the viral disease in question.

In addition, more drastic measures are sometimes followed, including destroying (roguing) infected crop and weed host plants and enforcing state and national quarantines or embargoes. Further control measures include controlling insect vectors by spraying plants with contact insecticides or fumigating soil to kill insects, nematodes, and other possible vectors. Growing valuable plants under fine cheesecloth or wire screening that excludes insect vectors also is done. Separation of new from virus-infected plantings of the same or closely related species is sometimes effective, and the simple practice of not propagating from plants suspected or known to harbour a virus also reduces loss.

Infected peach, apple, and rose budwood stock and carnations have been grown for weeks or months at temperatures about 37° to 38° C (99° to 100° F) to free new growth from viruses. Soaking some woody plant parts or virus- infected sugarcane shoots in hot water at about 50° C (120° F) for short periods also is effective. Both dry and wet heat treatments are based on the sensitivity of certain viruses to high temperatures. Rapidly growing dahlia and chrysanthemum sprouts outgrow viruses so that stem tips can be used to propagate healthy plants. With certain carnations, chrysanthemums, and potatoes, a few cells from the growing tip have been grown under sterile conditions in tissue culture; from these, whole plants have been developed free from viruses.

PRINCIPLES OF DISEASE CONTROL

Successful disease control requires thorough knowledge of the causal agent and the disease cycle, host-pathogen interactions in relation to environmental factors, and cost. Disease control starts with the best variety, seed, or planting stock available and continues throughout the life of the plant. For harvested crops, disease control extends through transport, storage, and marketing. Relatively few diseases are controlled by a single method; the majority require several approaches. These often need to be integrated into a broad programme of biological, cultural, and chemical methods to control as many different pests—including insects, mites, rodents, and weeds—on a given crop as possible. Most control measures are directed against inoculum of the pathogen and involve the principles of exclusion and avoidance, eradication, protection, host resistance and selection, and therapy.

Exclusion and Avoidance

The principle of exclusion and avoidance is to keep the pathogen away from the growing host plant. This practice commonly excludes pathogens by disinfection of plants, seeds, or other parts, using chemicals or heat. Inspection and certification of seed and other planting stock help ensure freedom from disease. For gardeners this involves sorting bulbs or corms before planting and rejecting diseased plants. Federal and state plant quarantines, or embargoes, have been established to prevent introduction of potentially destructive pathogens into areas currently free of the disease. More than 150 countries now have established quarantine regulations.

Eradication

Eradication is concerned with elimination of the disease agent after it has become established in the area of the growing host or has penetrated the host. Such measures include crop rotation, destruction of the diseased plants, elimination of alternate host plants, pruning, disinfection, and heat treatments.

Crop rotation with non-susceptible crops "starves out" bacteria, fungi, and nematodes with a restricted host range. Some pathogens can survive only as long as the host residue persists, usually no more than a year or two. Many pathogens, however, are relatively unaffected by rotation because they become established as saprophytes in the soil (e.g., Fusarium and Pythium species; Rhizoctonia solani; and the potato scab actinomycete, Streptomyces scabies) or their propagative structures remain dormant but viable for many years (e.g., cysts of cyst nematodes, sporangia of the cabbage clubroot fungus, and onion smut spores).

Burning, deep plowing of plant debris, and fall spraying are used against such diseases as leaf blights of tomato, Dutch elm disease, and apple scab. Destruction of weed hosts also helps control such viral diseases as cucumber mosaic and curly top. For fungi whose complete life cycle requires two different host species, such as black stem rust of cereals and white-pine blister rust, destruction of alternate hosts is effective. Destruction of diseased plants helps control Dutch elm disease, oak wilt, and peach viral diseases—mosaic, phony peach, and rosette. Elimination of citrus canker in the southeastern United States has been one of the few successful eradication programmes in history. Infected trees were sprayed with oil and burned.

Pruning and excision of a diseased portion of the plant have aided in reducing inoculum sources for canker and wood-rot diseases of shade trees and fire blight of pome fruits. Disinfection of contaminated tools, as well as packing and shipping containers, controls a wide range of diseases. Direct application of dry or wet heat is used to obtain seeds, bulbs, other propagative materials, and even entire plants free of viruses, nematodes, and other pathogens.

Protection

The principle of protection involves placing a barrier between the pathogen and the susceptible part of the host to shield the host from the pathogen. This can be accomplished by regulation of the environment, cultural and handling practices, control of insect carriers, and application of chemical pesticides.

Regulation of the Environment

Selection of outdoor growing areas where weather is unfavourable for disease is a method of controlling disease by regulating the environment. Control of viral diseases of potato, for example, can be accomplished by growing the seed crop in northern regions where low temperatures are unfavourable for the aphid carriers. Another environmental factor that can be brought under control is the storage and in-transit environment. A variety of postharvest diseases of potato, sweet potato, onion, cabbage, apple, pear, and other crops are controlled in storage and shipment by keeping humidity and temperature low and by reducing the quantity of ethylene and other natural gases in storage houses.

Cultural Practices

Selection of the best time and depth of seeding and planting is an effective cultural practice that reduces disease impact. Shallow planting of potatoes may help to prevent Rhizoctonia canker. Early fall seeding of winter wheat may be unfavourable for seedling infection by wheat-bunt teliospores. Cool- temperature crops can be grown in soils infested with root-knot nematode and harvested before soil temperatures become favourable for nematode activity. Adjustment of soil moisture is another cultural practice of widespread usefulness. For example, seed decay, damping-off (the destruction of seedlings at the soil line), and other seedling diseases are favoured by excessively wet soils. The presence of drain tiles in poorly drained fields and the use of ridges or beds for plants are often beneficial. Adjustment of soil pH also leads to control of some diseases. Common potato scab can be controlled by adjusting the pH to 5.2 or below; other acid-tolerant plants then must be used in crop rotation, however.

Regulation of Fertility Level and Nutrient Balance

Potash and nitrogen, and the balance between the two, may affect the incidence of certain bacterial, fungal, and viral diseases of corn, cotton, tobacco, and sugar beet. A number of microelements, including boron, iron, zinc, manganese, magnesium, copper, sulfur, and molybdenum, may cause non- infectious diseases of many crop and ornamental plants. Adjusting the soil pH, adding chelated (bound or enclosed in large organic molecules) or soluble salts to the soil, or spraying the foliage with these or similar salts is a corrective measure.

Handling Practices

Late blight on potato tubers can be controlled by delaying harvest until the foliage has been killed by frost, chemicals, or mechanical beaters. Avoidance of bruises and cuts while digging, grading, and packing potatoes, sweet potatoes, and bulb crops also reduces disease incidence.

Control of Insect Vectors

There are many examples in which losses by bacteria, viruses, and mycoplasma-like disease agents can be reduced by controlling aphids, leafhoppers, thrips, beetles, and other carriers of these agents.

Chemical Control

A variety of chemicals are available that have been designed to control plant diseases by inhibiting the growth of or by killing the disease-causing pathogens. Chemicals used to control bacteria (bactericides), fungi (fungicides), and nematodes (nematicides) may be applied to seeds, foliage, flowers, fruit, or soil.

They prevent or reduce infections by utilizing various principles of disease control. Eradicants are designed to kill a pathogen that may be present in the soil, on the seeds, or on vegetative propagative organs, such as bulbs, corms, and tubers. Protectants place a chemical barrier between the plant and the pathogen. Therapeutic chemicals are applied to combat an infection in progress. Soil treatments are designed to kill soil-inhabiting nematodes, fungi, and bacteria. This eradication can be accomplished using steam or chemical fumigants. Soilborne nematodes can be killed by applying granular or liquid nematicides. Most soil is treated well before planting; however, certain fungicides can be mixed with the soil at planting time.

Seeds, bulbs, corms, and tubers are frequently treated with chemicals to eradicate pathogenic bacteria, fungi, and nematodes and to protect the seeds against organisms in the soil—mainly fungi—that cause decay and damping- off. Seeds are often treated with systemic fungicides, which are absorbed and provide protection for the growing seedling.

Protective sprays and dusts applied to the foliage and fruit of crops and ornamentals include a wide range of organic chemicals designed to prevent infection. Protectants are not absorbed by or translocated through the plant; thus they protect only those parts of the plant treated before invasion by the pathogen. A second application is often necessary because the chemical may be removed by wind, rain, or irrigation or may be broken down by sunlight. New, untreated growth also is susceptible to infection. New chemicals are constantly being developed.

Biological Control

Biological control of plant diseases involves the use of organisms other than humans to reduce or prevent infection by a pathogen. These organisms are called antagonists; they may occur naturally within the host's environment, or they may be purposefully applied to those parts of the potential host plant where they can act directly or indirectly on the pathogen.

Although the effects of biological control have long been observed, the mechanisms by which antagonists achieve control is not completely understood. Several methods have been observed: some antagonists produce antibiotics that kill or reduce the number of closely related pathogens; some are parasites on pathogens; and others simply compete with pathogens for available food. Cultural practices that favour a naturally occurring antagonist and exploit its beneficial action often are effective in reducing disease. One technique is to incorporate green manure, such as alfalfa, into the soil. Saprophytic microorganisms feed on the green manure, depriving potential pathogens of available nitrogen. Another practice is to make use of suppressive soils—those in which a pathogen is known to persist but causes little damage to the crop. A likely explanation for this phenomenon is that suppressive soils harbour antagonists that compete with the pathogen for food and thereby limit the growth of the pathogen population.

Other antagonists produce substances that inhibit or kill potential pathogens occurring in close proximity. An example of this process, called antibiosis, is provided by marigold (Tagetes species) roots, which release terthienyls, chemicals that are toxic to several species of nematodes and fungi.

Only a few antagonists have been developed specifically for use in plant- disease control. Citrus trees are inoculated with an attenuated strain of tristeza virus, which effectively controls the virulent strain that causes the disease. An avirulent strain of Agrobacterium radiobacter (K84) can be applied to plant wounds to prevent crown gall caused by infection with Agrobacterium tumefaciens. Many more specific antagonists are being investigated and hold much promise for future control of disease.

Therapy

Therapeutic measures have been used much less often in plant pathology than in human or animal medicine. The recent development of systemic fungicides such as oxathiins, benzimidazoles, and pyrimidines have enabled growers to treat many plants after an infection has begun. Systemic chemicals are absorbed by and translocated within the plant, restricting the spread and development of pathogens by direct or indirect toxic effects or by increasing the ability of the host to resist infection.

Antibiotics have been developed to control various plant diseases. Most of these drugs are absorbed by and translocated throughout the plant, providing systemic therapy. Streptomycin is used against a variety of bacterial pathogens; tetracycline is able to control the growth of certain mycoplasmas; and cycloheximides offer effective control for certain diseases caused by fungi.

Host Resistance and Selection

Disease-resistant varieties of plants offer an effective, safe, and relatively inexpensive method of control for many crop diseases. Most available commercial varieties of crop plants bear resistance to at least one, and often several, pathogens. Resistant or immune varieties are critically important for low-value crops in which other controls are unavailable, or their expense makes them impractical. Much has been accomplished in developing disease-resistant varieties of field crops, vegetables, fruits, turf grasses, and ornamentals. Although great flexibility and potential for genetic change exist in most economically important plants, pathogens are also flexible. Sometimes, a new plant variety is developed that is highly susceptible to a previously unimportant pathogen.

Variable Resistance

Resistance to disease varies among plants; it may be either total (a plant is immune to a specific pathogen) or partial (a plant is tolerant to a pathogen, suffering minimal injury). The two broad categories of resistance to plant diseases are vertical (specific) and horizontal (non-specific). A plant variety that exhibits a high degree of resistance to a single race, or strain, of a pathogen is said to be vertically resistant; this ability usually is controlled by one or a few plant genes. Horizontal resistance, on the other hand, protects plant varieties against several strains of a pathogen, although the protection is not as complete. Horizontal resistance is more common and involves many genes.

Obtaining Disease-resistant Plants

Several means of obtaining disease-resistant plants are commonly employed alone or in combination. These include introduction from an outside source, selection, and induced variation. All three may be used at different stages in a continuous process; for example, varieties free from injurious insects or plant diseases may be introduced for comparison with local varieties. The more promising lines or strains are then selected for further propagation, and they are further improved by promoting as much variation as possible through hybridization or special treatment. Finally, selection of the plants showing greatest promise takes place. Developing disease-resistant plants is a continuing process.

Special treatments for inducing gene changes include the application of mutation-inducing chemicals and irradiation with ultraviolet light and X rays. These treatments commonly induce deleterious genetic changes, but, occasionally, beneficial ones also may occur.

Methods used in breeding plants for disease resistance are similar to those used in breeding for other characters except that two organisms are involved— the host plant and the pathogen. Thus, it is necessary to know as much as possible about the nature of inheritance of the resistant characters in the host plant and the existence of physiological races or strains of the pathogen.

The use of Genetic Engineering in Developing Disease-resistant Plants

The techniques of genetic engineering can be used to manipulate the genetic material of a cell in order to produce a new characteristic in an organism. Genes from plants, microbes, and animals can be recombined (recombinant DNA) and introduced into the living cells of any of these organisms. Organisms that have had genes from other species inserted into their genome (the full complement of an organism's genes) are called transgenic. The production of pathogen-resistant transgenic plants has been achieved by this method; certain genes are inserted into the plant's genome that confer resistance to such pathogens as viruses, fungi, and insects. Transgenic plants that are tolerant to herbicides and that show improvements in other qualities also have been developed.

Apprehension about the release of transgenic plants into the environment exists, and measures to safeguard the application of this technology have been adopted. In the United States several federal agencies, such as the U.S. Department of Agriculture, the Food and Drug Administration, and the Environmental Protection Agency, regulate the use of genetically engineered organisms. As of 2006, more than 250 million acres (100 million hectares) worldwide were planted with genetically modified (GM) crops. Among the most successful GM crops are corn (maize), soybeans, and cotton, all of which have proved valuable to farmers with respect to producing increased yields and having economic advantages.

BACTERIAL DISEASES OF BEANS

Many bacterial pathogens occur throughout the dry and snap bean growing areas of Colorado and surrounding states. Yield losses due to bacterial pathogens (including seed size and quality) may range from a trace to 100 per cent, especially when adverse environmental conditions persist during the early growth and flowering stages. In Colorado, halo blight (caused by Pseudomonas syringae pv. phaseolicola), bacterial brown spot (caused by Pseudomonas syringae pv. syringae) and common bacterial blight (caused by Xanthomonas campestris pv. phaseoli or Xanthomonas axonopodis pv. phaseoli) are the most common foliar, bacterial diseases of beans. Bacterial wilt (caused by Curtobacterium flaccumfaciens ssp. flaccumfaciens) has also been reported in our region.

PATHOGEN SURVIVAL

Bacterial pathogens can overwinter in previously infected bean debris (straw and seed) in old bean ground and as saprophytes on and in bean tissue. The following year, surviving bacteria can multiply on emerging, contaminated volunteer beans. They can also be spread from plant debris by wind, rain and irrigation water.

The bacteria may reside on the surface of bean leaves as epiphytes without causing disease, or may incite lesions under favourable environmental conditions. This primary inoculum can then be easily spread by wind and water into nearby bean fields, often resulting in subsequent disease outbreaks during favourable conditions.

Disease Development

The incidence and severity of bacterial diseases are affected greatly by environmental, host and cultural factors. Halo blight and bacterial brown spot disease outbreaks are most serious when temperatures are moderately cool (below 80 degrees F) and humidity is high (greater than 95 per cent) for 24 hours or longer. Typically these pathogens most commonly infect young foliage and symptoms are rarely seen on older leaves. Common bacterial blight and bacterial wilt are favoured by higher temperatures (greater than 80 degrees) usually prevalent during July and August. Symptoms usually appear on middle-aged to older leaves, even at low humidity.

Severe disease outbreaks often occur shortly (seven to ten days) after a period of humid, rainy weather. Violent storms with hail and high winds cause plant wounding which enable pathogens to enter and infect plant tissues. Wounds caused by damaging plants with farm machinery or field workers during cultivation can spread pathogens, especially when foliage is wet. Planting non-certified seed (which can be contaminated) also contributes to serious disease outbreaks due to the seed-borne nature of these pathogens.

Disease Symptoms

Halo blight symptoms on leaves initially appear as greasy, water-soaked spots about 1/16 inch in diameter. They are generally most visible on the underside of young leaflets approximately one week after infection. One or more water-soaked spots are later surrounded by a greenish-yellow halo about 1/16 to 1/2 inch in diameter. This halo is caused by a bacterial toxin. The size and development of this halo is variable, and may not be produced during high temperatures. During severe infection, the disease may become systemic and cause yellowing and death of new foliage. Halo blight symptoms on pods appear as small circular, water-soaked spots or streaks. A light cream or silver- coloured bacterial ooze is often associated with these spots. Pod infection often causes discolouration, shriveling and bacterial contamination of seeds; however, some seed may appear to be healthy.

Bacterial brown spot symptoms are similar to young halo blight lesions after initial infection. The symptoms appear as small water-soaked spots most visible on the underside of young foliage.

A narrow, greenish yellow border about 1/16 inch wide may surround some of the lesions. However, when the lesion matures, it typically develops a "brown spot" appearance, and dead tissue in the centre may fall out, producing a shot-hole appearance. Infected pods may be twisted and kinked and exhibit circular brownish water-soaked spots.

Common bacterial blight symptoms exhibit a scalded appearance on leaf tissue and contain water-soaked spots. These small, irregularly shaped lesions often enlarge to greater than 1 inch or more and form dark brown lesions along the edge of the leaflet.

A narrow lemon-yellow margin often surrounds these lesions and large portions of the foliage can be infected. Infected pods exhibit circular, water- soaked areas that often produce yellow masses of bacterial ooze. Later, spots dry and appear as reddish-brown lesions. Pod infection often causes discolouration, shriveling and bacterial contamination of seeds; however, some seed may appear healthy.

Bacterial wilt symptoms may appear after plugging of the water-conducting tissue of young seedlings which may wilt and die. Leaves of older infected plants will wilt, especially during moisture stress and warmer parts of the day. Golden brown, irregularly-shaped leaf lesions occur, with or without evidence of water-soaking, and affected leaves wilt and may drop off. Infection can occur on pod sutures similar to that caused by other bacterial diseases, but seldom will it produce circular water-soaked spots. The bacterium may cause a seed discolouration (yellow, orange, purple), depending on the strain of wilt organism and bean market class.

Disease Management

Bacterial pathogens of dry and snap beans can be managed, but not eliminated, by carefully implementing integrated crop production strategies. These approaches reduce or delay disease severity during the critical periods of vegetative and reproductive plant growth.

Overwintering populations can effectively be reduced by employing a three-to four-year rotation with beans planted once every third or fourth year. Corn, small grains and vegetables are recommended crop alternatives during two of these years. Proper sanitation of bean crop debris is important. It requires the complete incorporation of straw into the soil and the elimination of volunteer beans early the following year. This reduces diseased plants which provide the inoculum responsible for disease outbreaks in nearby bean fields. Bacterial pathogens can also survive in bean dust on contaminated harvest equipment, seed-cleaning equipment and storage containers.

Always plant high quality certified seed to minimize early season disease. Treat seeds with streptomycin, which effectively kills most external bacterial contamination. Diseases can still develop in these fields, but the pathogens must be transported into the field from external inoculum sources.

Avoid cultivation of beans when plants are wet or when the stand is too tall to allow machinery to pass through without wounding the foliage. Thoroughly clean soil, weeds and crop residues from equipment before moving to other fields. Reuse of irrigation run-off water is not recommended since pathogens are transmitted by it. Sprinkler irrigation can increase the spread and severity of bacterial foliar diseases by splashing bacteria from plant to plant, and extending leaf wetness periods. Planting resistant varieties is the best method to manage bacterial diseases of beans. Most older bean varieties, especially pintos and light red kidneys, are susceptible to these bacterial diseases. Fortunately, blight resistant varieties are available in other market classes. Most navy and small white varieties, including Aurora, are resistant to halo blight. Consult your local Colorado State University Extension and bean industry personnel regarding new resistant varieties for your area. In Colorado, copper-based bactericides have effectively reduced populations of pathogenic pseudomonads on bean foliage and reduced the spread of these pathogens in infected Crops. Common bacterial blight and bacterial wilt are not controlled as effectively by copper spray programmes, although some reduction in disease may be obtained with a preventive programme. Applying these protectants early in the season (June to July) every seven to 10 days during cool to warm, moist weather can decrease establishment of bacterial pathogens.

To ensure complete coverage, these products should be applied with at least 5 gallons of water per acre. When pathogens have become established and disease symptoms are evident, copper sprays can help reduce their spread to healthy foliage and pods. But bacteria inside lesions are not affected. Commonly used cupric hydroxide bactericides include Kocide, NuCop and Champ. Always read the product label carefully and consult extension and industry representatives for updated recommendations. Closely inspect bean fields for symptoms of bacterial diseases throughout the growing season, especially after prolonged periods of high humidity. Apply a bactericide every seven to 10 days until plants are within 2 to 3 weeks from harvest; continue protection if infection is detected and weather conditions remain favourable for disease development. Hail-damaged beans should receive a preventive application of a copper bactericide to protect damaged plants during regrowth.

It must be emphasized that to minimize the impact of bacterial diseases on bean Crops in our region the producer must carefully integrate recommended strategies of crop rotation, sanitation, use of treated certified seed, varietal selection, stress and wound avoidance, and proper pesticide scheduling.

❐

Chapter - 5

Classification of Microbes

A microbe is any living thing that spends its life at a size visible sometimes only with a microscope. It is too tiny to be seen with the naked eye. Microbes are the oldest form of life on Earth. Some types have existed for billions of years. They may live as individuals or cluster together in communities. Microbes live in the water we drink, the food we eat and the air we breathe.

Right now, billions of microbes are swimming in our belly and mouth and crawling on our skin! Don't worry, over 95 per cent of microbes are good for you. Microbes include bacteria, viruses, fungi, algae and protozoa. These single-cell organisms are invisible to the eye, but they can be seen with microscopes.

Fig. Microbe

The term microbe is short for microorganism, which means small organism. To help people understand the different types of microbes, they are grouped or classified in various ways. Microbes are very diverse and represent all the great kingdoms of life. In fact, in terms of numbers, most of the diversity of life on Earth is represented by microbes.

WHAT ARE MICROBES ?

Microbes are tiny organisms-too tiny to see without a microscope, yet they are abundant on Earth. They live everywhere-in air, soil, rock, and water. Some live happily in searing heat, while others thrive in freezing cold. Some microbes need oxygen to live, but others do not. These microscopic organisms are found in plants and animals as well as in the human body.

Some microbes cause disease in humans, plants, and animals. Others are essential for a healthy life, and we could not exist without them. Indeed, the relationship between microbes and humans is delicate and complex.

Most microbes belong to one of four major groups:bacteria, viruses, fungi, or protozoa. A common word for microbes that cause disease is "germs."Some people refer to disease-causing microbes as "bugs." "I've got the flu bug,"for example, is a phrase you may hear during the wintertime to describe an influenza virus infection.

We know about large plants and animals like trees and horses, and small ones like grass and mice, and very small ones like mosses and ants, but there are many other living things even smaller then that, so small that we cannot see them with the naked eye. They are called microbes. This whole world of tiny living things is very important in the environment, so we must learn to understand, appreciate and respect them.

There are many kinds of microbes, and they are everywhere around us, in the soil, in the air, on our skins and even inside of us. Some microbes are tiny plants calledalgae that usually live in the water or on land. They make their food with energy from the sun just like bigger plants. When they are very numerous, they may colour the water green or brown, or they may appear as slimes or scums.

The greenish black that often stains the outside of concrete buildings in the tropics comes from algae. Algae may be green, blue-green, brown, yellow- brown or even red or orange, like the alga that sometimes grows on the trunks of coconut palms near the sea. There are also tiny animals or protozoa of many kinds that swim or crawl around and eat things like algae and other animals. Since most of them are colourless, they are hard to see without some kind of magnification.

The bacteria or germs are even smaller than the algae, and there are millions and millions of them everywhere. They live by dissolving and breaking down any kind of organic matter, which includes anything that is or was part of a living thing. Thus they break down anything that is dead, and sometimes they do not wait for it to be dead, which is often how things get sick. It is usually bacteria that cause milk to go sour or food to spoil. Even though a surface may look clean, it has bacteria on it unless it has been sterilised (made clean of all bacteria) by heating, boiling or the application of alcohol or an antiseptic (a chemical that kills bacteria).

Another important group of microbes are the fungi, which grow like plants but get their food by breaking down dead material. They include the moulds and mildews which are responsible for rotting and decay. Mushrooms are the fruiting bodies of certain kinds of fungi. There are three important kinds of activities that microbes carry out in the environment:

DECOMPOSERS

These microbes are an essential part of all ecosystems, because they decompose or break down dead material and keep it from accumulating. If something did not break down the dead leaves in a forest, they would pile up higher and higher until the trees were buried under them. When the decomposers break things down, they release the materials of which these things were made, particularly the nutrients, which are then available to be taken up by plant roots and used again. If the nutrients were not released, they would eventually all be tied up in dead material, and then nothing else could grow. Sometimes the decomposers create a problem when they rot or spoil things faster than we would like.

SYMBIONTS

Sometimes microbes are actually helpful to plants and animals in a relationship called symbiosis because each one helps the other. For instance, there are certain fungi in the soil which grow attached to the roots of trees and which help the tree to absorb water and nutrients from the soil. This help is so important for some trees that they cannot grow without their special fungus.

Only certain bacteria and blue-green algae are able to fix nitrogen from the air and make it available as a nutrient to plants. In a few plants like the legumes, the roots have special swellings, called nodules, where nitrogen-fixing bacteria can live under good conditions inside the roots and make nitrogen nutrients for the plant.

Animals like corals and giant clams have one-celled algae (tiny plants) living inside their bodies; the animals give the plants shelter and nutrients and expose them to the light, and the plants make food for the animals. Then there are animals like cows and termites that have special bacteria living in their stomachs to help them digest food in the grass and wood that they could not use otherwise.

DISEASE GERMS

There are always many kinds of microbes on and in our bodies, but most of them do not hurt us, and a few are even helpful. Unfortunately, some microbes are much more dangerous, and want to live off of us or of other living things before we are dead. Some may even try to kill us to get more food. As they multiply in our bodies they make us sick. Our bodies have some defences against these disease germs, but sometimes they are not enough, and we need the help of medicines to overcome the germs and get well again.

Antibiotics, for instance, are special poisons against microbes, just as insecticides are poisons against insects. Doctors sometimes must give us these strong medicines to help our body get rid of a dangerous disease germ, but they must be used with care or they can upset the body's balance in other ways.

LIVING WITH MICROBES

In thinking about the environment, we must remember that the microbes are always there even if we cannot see them. Often they are useful, but sometimes they are dangerous, and we must learn how to be careful and control them. Microbes that cause disease are often spread from one person to another, making many people sick. When a person is sick, there may be many disease germs in his mouth or in his wastes. If he sneezes on his hand, there will be germs on his hand. If he touches a child, the sickness could pass to the child; if he touches a load of bread, the germs could pass to anyone who eats the bread.

Suppose a person has been sick in the bush. An animal might pick up his germs and bring them back to the village. Pigs, rats, cockroaches and flies, all may carry germs from dirty or unsanitary places and put them on food or in places where children play. This is why if a place is clean there is less chance of spreading sicknesses.

Germs are often spread through dirty or contaminated water. Imagine a bucket that everyone uses. It may be set on the ground where dirty pigs have been walking, or be used for washing dirty clothes or a sick babies' diapers, then it may be lowered into a well or cistern to get water for washing. The germs on the sides and bottom will get into the water, and then can be given to anyone drinking the water or eating food or from plates that have been washed in it. Since you cannot see germs, they may be there even if something looks clean.

There are things that can be done to control the harmful effects of germs or to keep them from spreading. Microbes are always waiting to break down anything that is lying around, like our food, for example. Since microbes usually grow faster when it is warm, food spoils faster (from attack by microbes) in warm weather. If the food is put in a cold place like a refrigerator, the microbes grow more slowly, and the food will keep several days.

The colder it is, the longer the food can be kept without spoiling. In a very cold freezer, the microbes can hardly grow at all. In the same way, enough heat can kill microbes. Boiling water for 20 minutes will kill the germs in it, or in any food put in it. The food in cans does not spoil because the cans were heated in the cannery to kill all the microbes inside. Boiling baby bottles and nipples, and any water put inside them, will sterilize them and kill all the germs, since babies are still weak at fighting germs and get sick very easily.

There are also chemicals that can be used as disinfectants to kill microbes, including chlorine bleach, alcohol, and the antiseptics used to treat cuts and other injuries which might otherwise let microbes into the body. Washing with soap or detergent can reduce but not eliminate microbes. If someone is sick and there is a danger of contamination, then a disinfectant can be used to keep the germs from being spread.

Since the 19th century, we have known microbes cause infectious diseases. Near the end of the 20th century, researchers began to learn that microbes also contribute to many chronic diseases and conditions. Mounting scientific evidence strongly links microbes to some forms of cancer, coronary artery disease, diabetes, multiple sclerosis, and chronic lung diseases.

Microbes are single-cell organisms so tiny that millions can fit into the eye of a needle. They are the oldest form of life on earth. Microbe fossils date back more than 3.5 billion years to a time when the Earth was covered with oceans that regularly reached the boiling point, hundreds of millions of years before dinosaurs roamed the earth. Without microbes, we couldn't eat or breathe. Without us, they'd probably be just fine. There are more of them on a person's hand than there are people on the entire planet! Microbes are in the air we breathe, the ground we walk on, the food we eat—they're even inside us! We couldn't digest food without them—animals couldn't, either. Without microbes, plants couldn't grow, garbage wouldn't decay and there would be a lot less oxygen to breathe. In fact, without these invisible companions, our planet wouldn't survive as we know it!

- Our Challenge: How are we going to solve the case of what a microbe is when they're so small they're invisible to us?

Microbe is a term for tiny creatures that individually are too small to be seen with the unaided eye. Microbes include bacteria (back-tear-ee-uh), archaea (are-key-uh), fungi(fun-jeye) and protists (pro-tists). You've probably heard of bacteria and fungi before. Archaea are bacteria-like creatures that have some traits not found in any true bacteria. Protists include primitive algae (al-gee), amoebas (ah-me-buhs), slime molds and protozoa (pro-toe-zoh-uh). We can also include viruses (vye-rus-is) as a major type of microbe, though there is a debate as to whether viruses can be considered living creatures or not.

To solve the case of what a microbe is, we have to use tools such as high- power microscopes. Let's zoom in on some microbes and see what a few of these strange creatures look like.

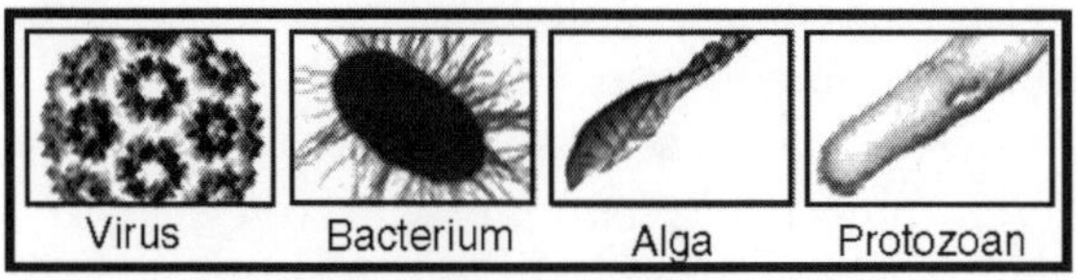

As you can see, microbes come in many varieties. They may live as individuals or cluster together in communities.

So how Small are Microbes

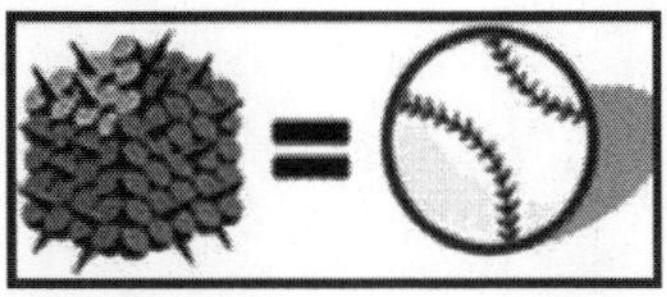

Well, let's say we could enlarge an average virus, the smallest of all microbes, to the size of a baseball.

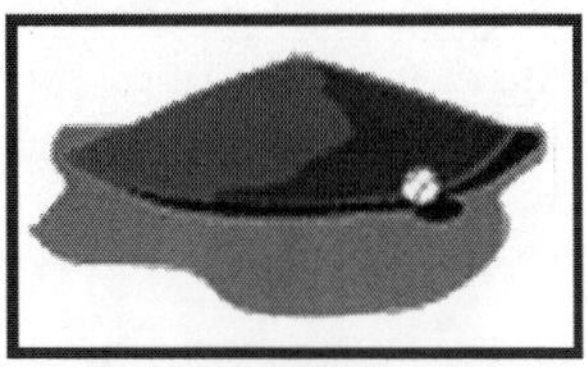

An average bacterium would then be the size of the pitcher's mound.

And just one of the millions of cells that make up your body would be the size of the ballpark!

TRANSMISSION

According to health care experts, infectious diseases caused by microbes are responsible for more deaths worldwide than any other single cause. They estimate the annual cost of medical care for treating infectious diseases in the United States alone is about $120 billion.

The science of microbiology explores how microbes work and how to control them. It seeks ways to use that knowledge to prevent and treat the diseases microbes cause. The 20th century saw an extraordinary increase in our knowledge about microbes. Microbiologists and other researchers had many successes in learning how microbes cause certain infectious diseases and how to combat those microbes. Unfortunately, microbes are much better at adapting to new environments than are people. Having existed on Earth for billions of years, microbes are constantly challenging human newcomers with ingenious new survival tactics.

- Many microbes are developing new properties to resist drug treatments that once effectively destroyed them. Drug resistance has become a serious problem worldwide.
- Changes in the environment have put certain human populations in contact with newly identified microbes that cause diseases we have never seen before, or that previously occurred only in isolated populations.
- Newly emerging diseases are a growing global health concern. Since 1976, scientists have identified approximately 30 new pathogens, such as HIV.

How you can be Protected from Microbes

We become immune to germs through natural and artificial means. As long ago as the 5th century B.C., Greek doctors noticed that people who had recovered from the plague would never get it again—they seemed to have become immune or resistant to the germ. The first time T cells and B cells in your immune system meet up with an antigen, such as a virus or bacterium, they prepare the immune system to destroy the antigen.

Naturally Acquired Immunity

Because the immune system often can remember its enemies, those cells become active if they meet that particular antigen again. This is called naturally acquired immunity.

Another example of naturally acquired immunity occurs when a pregnant woman passes antibodies to her unborn baby. Babies are born with weak immune responses, but they are protected from some diseases for their first few months of life by antibodies received from their mothers before birth. Babies who are nursed also receive antibodies from breast milk that help protect their digestive tracts.

Artificial Immunity

Artificial immunity can come from vaccines. Immunization with vaccines is a safe way to get protection from germs. Some vaccines contain microorganisms or parts of microorganisms that have been weakened or killed. If you get this type of vaccine, those microorganisms (or their parts) will start your body's immune response, which will demolish the foreign invader but not make you sick. This is a type of artificially acquired immunity.

Immunity can be strong or weak and short- or long-lived, depending on the type of antigen, the amount of antigen, and the route by which it enters your body. When faced with the same antigen, some people's immune system will respond forcefully, others feebly, and some not at all.

The genes you inherit also can influence your likelihood of getting a disease. In simple terms, the genes you get from your parents can influence how your body reacts to certain microbes.

Some Ways Microbes are Transmitted

You can transmit microbes to another person through the air by coughing or sneezing. These are common ways to get viruses that cause colds or flu, or the bacteria that cause tuberculosis (TB). Interestingly, international airplane travel can expose you to germs not common in your own country.

Close Contact can Pass Germs to Another Person

Scientists have identified more than 500 types of bacteria that live in our mouths. Some keep the oral environment healthy, while others cause problems like gum disease. One way you can transmit oral bacteria is by kissing. Microbes such as HIV, herpes simplex virus type 2, which causes genital herpes, and Neisseria gonorrhoeae bacteria, which causes gonorrhea, are examples of germs that cyou can pass to another during sexual intercourse.

You can Pick Up and Spread Germs by Touching Infectious Material

A common way for some microbes to enter the body, especially when caring for young children, is through unintentionally passing feces from hand to mouth or the mouths of young children. Infant diarrhea is often spread in this way. Daycare workers, for example, can pass diarrhea-causing rotavirus or Giardia lamblia (protozoa) from one baby to the next between diaper changes and other childcare practices. It also is possible to pick up cold viruses from shaking someone's hand or from touching contaminated surfaces, such as a doorknob or computer keyboard.

A Healthy Person can Carry Germs and Pass them onto Others

The story of "Typhoid Mary" is a famous example from medical history about how a person can pass germs on to others, yet not be affected by those germs.

The germs in this case were Salmonella typhi bacteria, which cause typhoid fever and are usually spread through food or water. In the early 20th century,

Mary Mallon, an Irish immigrant, worked as a cook for several New York City families. More than half of the first family she worked for came down with typhoid fever. Through a clever deduction, a researcher determined that the disease was caused by the family cook. He concluded that although Mary had no symptoms of the disease, she probably had had a mild typhoid infection sometime in the past. Though not sick, she still carried the Salmonella bacteria and was able to spread them to others through the food she prepared.

Germs from your Household Pet can Make you Sick

You can catch a variety of germs from animals, especially household pets. The rabies virus, which can infect cats and dogs, is one of the most serious and deadly of these microbes. Fortunately, rabies vaccine prevents animals from getting rabies. Vaccines also protect people from accidentally getting the virus from an animal. In addition, vaccines prevent people who already have been exposed to the virus, such as through an animal bite, from getting sick.

Dog and cat saliva can contain any of more than 100 different germs that can make you sick. Pasteurella bacteria, the most common, can be transmitted through bites that break the skin causing serious, and sometimes fatal, diseases such as meningitis. Meningitis is the inflammation of the lining of the brain and spinal cord. Warm-blooded animals are not the only ones that can cause you harm. Pet reptiles such as turtles, snakes, and iguanas can give Salmonella bacteria to their unsuspecting owners.

You can Get Microbes from Tiny Critters

Mosquitoes may be the most common carriers, also called vectors, of pathogens. Anopheles mosquitoes can pick up Plasmodium, which causes malaria, from the blood of an infected person and transmit the protozoan to an uninfected person. Fleas are more closely related to crabs than to insects, are another common vector. The tiny deer tick can infect humans with Borrelia burgdorferi, the bacterium that causes Lyme disease, which the tick picks up from mice.

Some Microbes in Food or Water could Make you Sick

Every year, millions of people worldwide become ill from eating contaminated foods. Although many cases of foodborne illness or "food poisoning" are not reported, the Centres for Disease Control and Prevention (CDC) estimates there are 76 million cases of such illnesses in the United States each year. In addition, CDC estimates 325,000 hospitalizations and 5,000 deaths are related to foodborne diseases each year. Microbes can cause these illnesses, some of which can be fatal if not treated properly. Poor manufacturing processes or poor food preparation can allow microbes to grow in food and subsequently infect you. Escherichia coli (E. coli) bacteria sometimes persist in food products such as undercooked hamburger meat and unpasteurized fruit juice. These bacteria can have deadly consequences in vulnerable people, especially children and the elderly. Cryptosporidia are bacteria found in human and animal feces. These bacteria can get into lake, river, and ocean water from sewage spills, animal waste, and water run-off. Millions can be released from infectious fecal matter. People who drink, swim in, or play in infected water can get sick. People, including babies, with diarrhea caused by Cryptosporidia or other diarrhea-causing microbes such as Giardia and Salmonella, can infect others while using swimming pools, waterparks, hot tubs, and spas.

Transplanted Animal Organs may Harbour Germs

Researchers are investigating the possibility of transplanting animal organs, such as pig hearts, into people. They, however, must guard against the risk that those organs also may transmit microbes that were harmless to the animal into humans, where they may cause disease.

SYMPTOMS

Generally, you should consult your health care provider if you have or think you may have an infectious disease. These trained professionals can determine whether you have been infected, determine the seriousness of your infection, and give you the best advice for treating or preventing disease. Sometimes, however, a visit to the doctor may not be necessary. Some infectious diseases, such as the common cold, usually do not require a visit to your doctor. They often last a short time and are not life-threatening, or there is no specific treatment. We've all heard the advice to rest and drink plenty of liquids to treat colds. Unless there are complications, most victims of colds find that their immune systems successfully fight off the viral culprits. In fact, the coughing and sneezing that make you feel miserable are part of your immune system's way of fighting off the culprits.

If, however, you have other conditions in which your immune system doesn't function properly, you should be in contact with your health care provider whenever you suspect you have any infectious disease, even the common cold. Such conditions can include asthma and immune deficiency diseases like HIV/AIDS.

In addition, some common, usually mild infectious diseases, such as chickenpox or seasonal flu, can cause serious harm in very young children and the elderly.

You should call a health care provider immediately if…

- You have been bitten by an animal
- You are having difficulty breathing
- You have a cough that has lasted for more than a week
- You have a fever higher than 100°F
- You have episodes of rapid heartbeat
- You have a rash (especially if you have a fever at the same time)
- You have swelling
- You suddenly start having difficulty with seeing (blurry vision, for example)
- You have been vomiting

DIAGNOSIS

Sometimes your health care provider can diagnose an infectious disease by listening to your medical history and doing a physical exam. For example, listening to you describe what happened and any symptoms you have noticed plays an important part in helping your doctor find out what's wrong.

Blood and urine tests are other ways to diagnose an infection. A laboratory expert can sometimes see the offending microbe in a sample of blood or urine viewed under a microscope. One or both of these tests may be the only way to determine what caused the infection, or they may be used to confirm a diagnosis.

In another type of test, your health care provider will take a sample of blood or other body fluid, such as vaginal secretion, and then put it into a special container called a Petri dish to see if any microbe "grows." This test is called a culture. Lab workers usually can identify certain bacteria, such as chlamydia, and viruses, such as herpes simplex, using this method.

X-rays, scans, and biopsies (taking a tiny sample of tissue from the infected area and inspecting it under a microscope) are among other tools your health care provider can use to make an accurate diagnosis.

All of the above procedures are relatively safe, and some can be done in your health care provider's office or a clinic. Others pose a higher risk to you because they involve procedures that go inside your body. One such invasive procedure is taking a biopsy from an internal organ. For example, one way a doctor can diagnose Pneumocystis carinii pneumonia, a lung disease caused by a fungus, is by doing a biopsy on lung tissue and then examining the sample under a microscope.

TREATMENT

How an infectious disease is treated depends on the microbe that caused it and sometimes on the age and medical condition of the person affected. Certain diseases are not treated at all, but are allowed to run their course, with the immune system doing its job alone. Some diseases, such as the common cold, are treated only to relieve the symptoms. Others, such as strep throat, are treated to destroy the offending microbe as well as to relieve symptoms.

CLASSIFICATION OF MICROBES

The term microbe is short for microorganism, which means small organism. To help people understand the different types of microbes, they are grouped or classified in various ways. Microbes are very diverse and represent all the great kingdoms of life. In fact, in terms of numbers, most of the diversity of life on Earth is represented by microbes.

Here is an outline of the major groups of microorganisms:

- Viruses
- Bacteria
- Algae
- Fungi
- Protozoa

Viruses

A virus is too small to be seen without a microscope. A virus is basically a tiny bundle of genetic material carried in a shell called the viral coat. Some viruses have an additional layer around this coat called an envelope. That's basically all there is to viruses.

There are thousands of different viruses that come in many shapes. Many are multi-sided or polyhedral. If you've ever looked closely at a cut gem, like the diamond in an engagement ring, you've seen an example of a polyhedral shape.

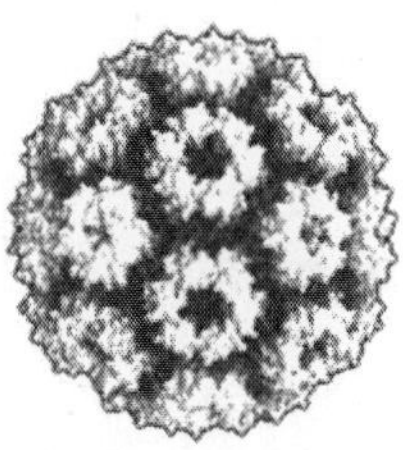

Fig. Viruses

Unlike the diamond in a ring, however, a virus does not taper to a point, but is shaped the same all around. Other viruses are shaped like spiky ovals or bricks with rounded corners. Some are like skinny sticks while others look like pieces of looped string. Some are more complex and shaped like little spaceship landing pods.

Viruses are found on or in just about every material and environment on Earth from soil to water to air. They're basically found anywhere there are cells to infect. Viruses can infect every living thing. However, viruses tend to be somewhat picky about what type of cells they infect. Plant viruses are not equipped to infect animal cells, for example, though a certain plant virus could infect a number of related plants. Sometimes, a virus may infect one animal and do no harm, but cause a great deal of damage when it gets into a different but closely related animal.

Viruses exist to reproduce only. To do that, they have to take over suitable host cells. Upon landing on a suitable host cell, a virus gets its genes inside the cell either by tricking the host cell to pull it inside, or by connecting its viral coat with the host cell wall or membrane and releasing its genes inside, or by injecting their genes into the host cell's DNA. The viral genes are then copied many times, using the process the host cell would normally use to reproduce its own DNA. The new viral genes then come together and assemble into whole new viruses. The new viruses are either released from the host cell without destroying the cell or eventually build up to a large enough number that they burst the host cell.

Bacteria

Bacteria consist of only one cell, but they're a very complex group of living things. Some bacteria can live in temperatures above the boiling point and in cold below the freezing point.

There are thousands of species of bacteria, but all of them are basically one of three different shapes. Some are rod- or stick-shaped; others are shaped like little balls. Others still are helical or spiral in shape. Some bacteria cells exist as individuals while others cluster together to form pairs, chains, squares or other groupings. Some bacteria can make their own food from sunlight, just like plants. Also like plants, they give off oxygen. Other bacteria absorb food from the material they live on or in. Some of these bacteria can live off iron or sulfur! The bacteria that live in our stomach absorb nutrients from the digested food you've eaten.

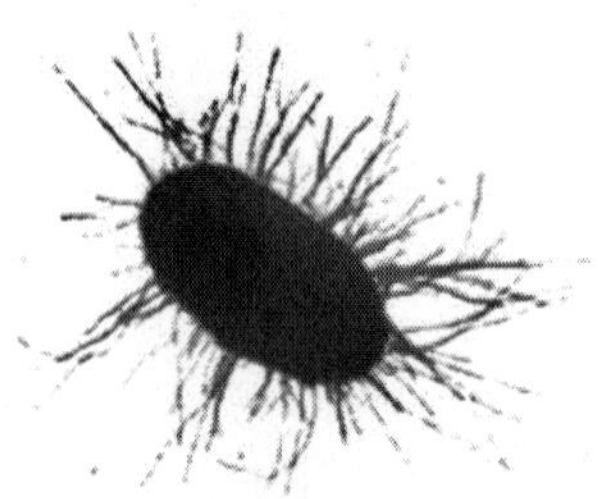

Fig. Bacteria

Some bacteria move about their environment by means of long, whip-like structures called flagella. They rotate their flagella like tiny outboard motors to propel themselves through liquid environments. They may also reverse the direction in which their flagella rotate so that they tumble about in one place. Other bacteria secrete a slime layer and ooze over surfaces like slugs. Others stay almost in the same spot. Bacteria live on or in just about every material and environment on Earth from soil to water to air and from our body to the Arctic ice to the Sahara deserts. Each square centimeter of our skin averages about 100,000 bacteria. A single teaspoon of soil contains more than a billion (1,000,000,000) bacteria.

Algae

Algae are found in fresh and salt water around the world. They can also grow on rocks and trees and in soil when enough water is available.

Most algae are able to make energy from sunlight, like plants do. They produce a large amount of the oxygen we breathe. However, at some stages of their lives, some algae get their nutrients from other living things.

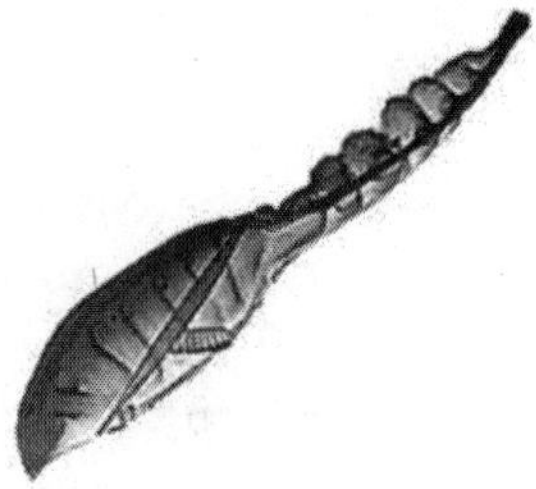

Fig. Algae

Diatoms are one kind of algae. They have hard shells made out of glass. When they die, these shells sink to the bottom of their watery environments. We mine deposits of these glass shells that formed hundreds of thousands of years ago to make abrasives, shiny road paint and grit in toothpaste. Diatoms come in all sorts of shapes—some are round and others are oval. Some look like leaves and others like fat commas.

Because algae make so much oxygen, these microbes are very helpful. But sometimes certain kinds of algae can also grow in such large numbers that when they suddenly die off en masse, the breaking down of their cells by bacteria destroys the amount of dissolved oxygen in the water, hurting the animals and plants that live there.

Fungi

Fungi are organisms that scientists once confused with plants. However, scientists have found that, at the cell level, the fungi are more like animals than they are like plants. For one thing, fungi cannot synthesize their own food like plants do, but instead they eat other organisms as do animals. Fungi come in a variety of shapes and sizes and different types. They can range from single cells to enormous chains of cells that can stretch for miles.

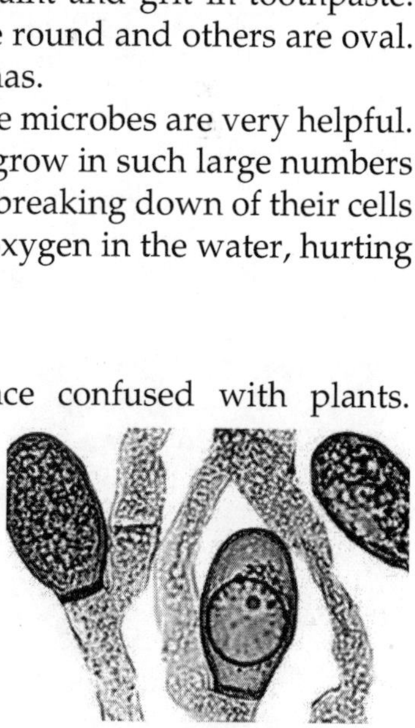

Fig. Fungi

Fungi include single-celled living things that exist individually, such as yeast and multicellular clusters, such as molds or mushrooms. Yeast cells look round or oval under a microscope. They're too small to see as individuals, but we can see large clusters of them as a white powdery coating on fruits and leaves.

Molds are described as filament-like because they form long filament-like, or thread-like, strands of cells called hyphae. These hyphae are what give mold colonies their fuzzy appearance. They also form the fleshy body, or mushroom, that some species grow. It may seem strange to think of something as big as a mushroom as a microbe. But the cells of the hyphae making up that mushroom are connected in a closer way than the cells of other multicellular living things, like we and me, are.

Fungi usually grow best in environments that are slightly acidic. They can grow on substances with very low moisture. Fungi live in the soil and on our body, in our house and on plants and animals, in freshwater and seawater. A single teaspoon of topsoil contains about 120,000 fungi.

Fungi are basically stationary. But they can spread either by forming reproductive spores that are carried on wind and rain or by growing and extending their hyphae.

Hyphae grow as new cells form at the tips, creating even longer chains of cells. Fungi absorb nutrients from living or dead organic matter that they grow on. They absorb simple, easily dissolved nutrients, such as sugars, through their cell walls. They give off special digestive enzymes to break down complex nutrients into simpler forms that they can absorb.

Some fungi are quite useful to us. We've used several kinds to make antibiotics to fight bacterial infections. These antibiotics are based on natural compounds the fungi produce to compete against bacteria for nutrients and space. We use baker's yeast, to make bread rise and to brew beer. Fungi break down dead plants and animals and keep the world tidier. We're exploring ways to use natural fungal enemies of insect pests to get rid of these bugs.

There are some dangerous fungi that cause diseases in plants, animals and people. Fungi ruin about a quarter to half of harvested fruits and vegetables each year.

IMPORTANT MICROBES

BACTERIA

Bacteria are found everywhere in water, soil, and even air. These small prokaryotic cells, typically from 0.2 to 1 um in length, are capable of living in boiling water, frozen ground, acid volcanoes, and at the bottom of the ocean. They can reproduce by doubling with a generation time of 20 minutes, or survive for centuries in a resting stage. In natural waters (lakes, streams, oceans) their generation time is around 1 day. In soils they live in a film of water around plant roots or other particles, and their activity is dependent on the temperature and the amount of available moisture. In general, bacteria are found in concentrations of 106 cells/mL of water in surface waters, and 109 cells/mL of soil in soils and sediments.

Some bacteria are capable of locomotion, and they possess the only rotary motor known in all of biology. This motor, similar to a wheel and axle, is capable of spinning a flagellum at speeds of 100 revolutions per second, or 6,000 rpm. Bacteria can propel themselves at a rate of 10 times their body length each second.

Bacteria, like all cells, are composed mostly of carbon, oxygen, nitrogen, hydrogen, phosphorus, and sulfur in the following percentages:

Element	% of dry weight
C	55
O	20
N	10
H	8
P	3
S	1

Bacteria take these elements and arrange them into polymers in the cells in the following percentages:

- 52.4% protein (amino acids, CHNOS)
- 19.9 per cent nucleic acid (organic bases, CNOHP)
- 16.6 per cent polysaccharide (sugar, CHO)
- 9.4 per cent phospholipid (C-16 acid + P, CHOP).

FUNGI

Fungi grow in the form of a finely-branched network of strands called hyphae which are 5-10 um in diameter. These hyphae can release digestive enzymes and take up nutrients over their entire length.

Fungi can absorb only small molecules such as sugars or peptides less than size amino acids. The reproductive organs of the fungi are called fruiting bodies or sporangia, which are sacs or other tissues that contain the fungi spores.

Fungi are uncommon in aquatic environments. On land, the amount of hyphae in the soil is measured in hundreds or thousands of meters of length per gram of soil. For example, the total length of hyphae in a gram of soil (about the amount that would fit on the fingernail of your little finger) can reach up to 1,600 meters (think about that for a minute).

Fungi secrete enzymes that can break down cellulose into glucose, one of the few kinds of organisms able to do this. Fungi are the only known organisms that degrade lignin completely. Cellulose and lignin are structural materials in plants that are difficult to degrade. The fungi do not use the breakdown products of lignin, but instead they use hydrogen peroxide to oxidize lignin in place. The breakdown products diffuse away, exposing the cellulose to enzymatic attack.

PROTOZOA

Protozoans are single-celled eukaryotes, not photosynthetic, that move by flagella or cilia. In oceans and lakes, the small 2-10 um long flagellates are the most important predators on bacteria. The larger ciliates (e.g., Paramecium) prey mostly upon photosynthetic cyanobacteria and small eukaryotic algae. In some termites, anaerobic protozoans in the gut degrade cellulose.

HUMAN EFFECTED BY MICROBES

"Ancient" diseases continue to be a problem where nutrition and sanitation are poor, and emerging diseases such as Acquired Immuno-deficiency Syndrome (AIDS) are even more dangerous for such populations. The Centers for Disease Control and Prevention (the U.S. government agency charged with protecting the health and safety of people) estimates that about 9 per cent of adults between the ages of 18-49 in Sub-Saharan Africa are infected with HIV. AIDS is only one of a number of new diseases that have emerged.

Many of these diseases have no known cure. Influenza and pneumonia are leading killers of the elderly even in the U.S. and other developed nations. Even the common cold causes illness and misery for almost everyone and drains the productivity of all nations. Many of the new diseases are viral in nature, making them notoriously difficult to treat. Disease due to food-borne pathogens also remains a problem, largely because of consumption of improperly processed or stored foods.

Understanding the sources of contamination and developing ways to limit the growth of pathogens in food is the job of food microbiologists. New infections continually appear.Having an available food source to grow on (humans) inevitably results in a microorganism that will take advantage. Surprisingly, many diseases that were previously thought to have only behavioural or genetic components have been found to involve microorganisms. The clearest case is that of ulcers, which was long thought to be caused by stress and poor diet. However the causative agent is actually a bacterium, Helicobacter pylori, and many ulcers can be cured with appropriate antibiotics.

Work on other non-infectious diseases such as heart disease, stroke and some autoimmune diseases also suggest a microbial component that triggers the illness. Finally, some pathogenic microbes that had been "controlled" through the use of antibiotics are beginning to develop drug resistance and therefore reemerge as serious threats in the industrialized world as well as developing nations. Tuberculosis is an illness that was on the decline until the middle 80's. It has recently become more of a problem, partly due to drug resistance and partly due to a higher population of immunosuppressed individuals from the AIDS epidemic.

Staphylococcus aureus strains are emerging that are resistant to many of the antibiotics that were previously effective against them. These staph infections are of great concern in hospital settings around the world. Understanding both familiar killers and new pathogens will require an understanding of their biology, and thus an understanding of the field of microbiology.

BENEFITS OF MICROBES

Significant resources have been spent to understand and fight disease-causing microorganisms from the beginning of microbiology. You may be surprised to learn that only a small faction of microbes are involved in disease; many other microbes actually enhance our well-being.

The harmless microbes that live in our intestines and on our skin actually help us fight off disease. They actively antagonize other bacteria and take up space, preventing potential pathogens from gaining a foothold on our bodies.

Indeed, like all large organisms, humans have entire communities of microorganisms in their digestive systems that contribute to their overall health. The microbial community in humans not only protects us from disease, but also provides needed vitamins, such as B_{12}. Human health and nutrition also depends on healthy farm animals. Cows, sheep, horses and other ruminant animals utilize their microbial associates to degrade plant material into useful nutrients.

Without these bacteria inside ruminants, growth on plant material would be impossible. In contrast to humans ruminant animals have complex stomachs that harbour large numbers of microorganisms. These microbes degrade the plant stuff eaten by the animal into usable nutrients. Without the assistance of the microbes, ruminant animals would not be able to digest the food they eat. Commercial crops are also central to human prosperity, and much of agriculture depends upon the activities of microbes. For example, an entire group of plants, the legumes, forms a cooperative relationship with certain bacteria.

ENVIRONMENT EFFECTED BY MICROBES

The vast majority of life on this planet is microscopic. These teaming multitudes profoundly influence the make-up and character of the environment in which we live. Presently, we know very little about the microbes that live in the world around us because less than 2 per cent of them can be grown in the laboratory. Understanding which microbes are in each ecological niche and what they are doing there is critical for our understanding of the world. Microbes are the major actors in the synthesis and degradation of all sorts of important molecules in environments.

Cyanobacteria and algae in the oceans are responsible for the majority of photosynthesis on Earth. They are the ultimate source of food for most ocean creatures (including whales) and replenish the world's oxygen supply. Cyanobacteria also use carbon dioxide to synthesize all of their biological molecules and thus remove it from the atmosphere.

Since carbon dioxide is a major greenhouse gas, its removal by cyanobacteria affects the global carbon dioxide balance and may be an important mitigating factor in global warming. In all habitats, microorganisms make nutrients available for the future growth of other living things by degrading dead organisms.

Microbes are also essential in treating the large volume of sewage and wastewater produced by metropolitan areas, recycling it into clean water that can be safely discharged into the environment.

Less helpfully (from the view of most humans), termites contain microorganisms in their guts that assist in the digestion of wood, allowing the termites to extract nutrients from what would otherwise be indigestible. Understanding of these systems helps us to manage them responsibly and as we learn more we will become ever more effective stewards. Energy is essential for our industrial society and microbes are important players in its production.

A significant portion of natural gas comes from the past action of methanogens (methane-producing bacteria). Numerous bacteria are also capable of rapidly degrading oil in the presence of air and special precautions have to be taken during the drilling, transport and storage of oil to minimize their impact. In the future, microbes may find utility in the direct production of energy. For example, many landfills and sewage treatment plants capture the methane produced by methanogens to power turbines that produce electricity.

Excess grain, crop waste and animal waste can be used as nutrients for microbes that ferment this biomass into methanol or ethanol. These biofuels are presently added to gasoline and thus decreasing pollution. They may one- day power fuel cells in our cars, causing little pollution and having water as their only emission.

Finally, We are increasingly taking advantage of the versatile appetite of bacteria to clean up environments that we have contaminated with crude oil, polycholrinated biphenyls (PCBs) and many other industrial wastes. This process is termed bioremediation and is a cheap and increasingly effective way of cleaning up pollution.

MICROBES IN GROUNDWATER

More than 95 per cent of the world's available fresh water (excluding ice caps and glaciers) is underground. This groundwater is valuable as a source of drinking water for most communities in the world, especially small ones.

In the United States, about 15 million private wells serve fewer than 25 people each. In addition, about 92 per cent of the 168,000 water systems serving 25 people or more get most or all of their drinking water from groundwater. The common microbes in groundwater (bacteria, protozoa, and viruses), primarily from a public health point of view.

COMMONLY OCCURRING MICROBES

Groundwater near the land surface usually teems with microbial life. Bacteria, which are microscopic single-celled organisms that lack a true nucleus and normally have a cell wall, are far more numerous than any other organism in the soil and groundwater. Near the surface where plant roots are abundant, there may be 100 million to 1 billion bacteria per gram of dry soil.

These values decrease dramatically with soil depth below the root zone, with densities depending on the amount of nutrients (food) and water available. Although concentrations of microbes below the root zone are lower than in the root zone itself, as many as 10 to 100 million bacteria per gram of aquifer material may be present. Bacteria have been found in core samples from a depth of 2.8 kilometers (1.7 miles) below the Earth's surface, and at a depth of 3.2 kilometers in South African gold mines. Protozoa, single-celled organisms that have a nucleus but lack a cell wall, also are common in groundwater. Protozoa typically are much larger than bacteria, and many types in groundwater feed on bacteria. Molds and other fungi also are common in groundwater near the land surface, where plenty of oxygen is present.

These microbes are larger than bacteria and, unlike the protozoa, have a cell wall and often grow in long filaments. Most fungi feed on dead or decaying material. Algae may even be present despite the absence of sunlight. Similar to the bacteria, the numbers of protozoa, fungi, and algae decrease with depth.

A large variety of viruses probably are also common in groundwater. Viruses are much smaller than other microbes and cannot be seen with ordinary microscopes. They do not have a nucleus or cell wall, and can multiply only within the cells of larger organisms. The natural viruses in groundwater are able to reproduce only by infecting and usually killing the bacteria and other larger microbes present.

Microbial Habitats

Microbes below Earth's surface must compete vigourously with each other for limited food and space. Bacteria that are not adapted to this habitat cannot survive long. Normally, nutrients are scarce in this environment, and thus microbes in groundwater grow slowly.

Because the microbes in groundwater and other environmental habitats are well adapted to life in these habitats, few of them have the ability or the need to infect and cause illness in humans and other animals. However, a few environmental bacteria and protozoa may cause disease in humans by chance under certain circumstances, especially in individuals with a weakened immune system or major breaks in the skin that allow microbes to enter.

Many naturally-occurring bacteria below the surface may serve a beneficial role by destroying human-made toxic chemicals that migrate from the surface to groundwater. Some toxic chemicals, such as gasoline, are easy for subsurface bacteria to break down, but others, such as dry-cleaning fluid and some pesticides, are destroyed only slowly, if at all.

Disease-Causing Pathways

The groundwater in a drinking-water well may contain a wide variety of microbes without presenting a public health risk. However, groundwater in some areas becomes contaminated by the fecal material of humans and other animals.

This is a cause for concern because fecal material may contain pathogenic (disease-causing) microbes that can infect the intestinal tract of humans. Fecal pathogens may be bacterial, viral, or protozoan. Water containing fecal material may seep into the groundwater from the land surface or from underground sources of contamination. Major surface sources include:

- Wastewater and biosolids from sewage treatment facilities that have been applied to land as a soil conditioner.
- Seepage from shallow artificial ponds (lagoons) used for processing sewage.
- Seepage from contaminated lakes and other surface-water bodies.
- Urban run-off.
- Feces from cattle and other livestock operations.
- Improperly constructed sanitary landfills where trash and garbage are disposed.

Fecal contamination also can reach the groundwater from underground sources, such as improperly functioning septic tank systems, underground reservoirs for liquid household sewage (cesspools), or leaking underground sewer lines. About 25 million septic tanks exist in the United States.

PATHOGEN MOVEMENT AND PERSISTENCE

Most pathogens from fecal matter remain either near the surface, or near the point of origin in the case of an underground source. However, where conditions are favourable, some pathogens travel along with the water flow though pores (tiny openings) in the surrounding soil and rock, and may enter groundwater.

Whether a pathogen reaches the groundwater, especially in the area of a well intake, depends on how strongly it is retained by the soil and how long it survives. The extent to which a pathogen is retained by the soil depends on:

- Soil characteristics, such as the number and size of pores and their interconnections, and how strongly the pathogen adsorbs (binds) to the soil particles;
- Environmental factors, especially the amount of rainfall; and
- The characteristics of an organism, including its size and its tendency to attach to other organisms and organic debris.

Rapid pathogen transport may occur in aquifers that have large pores (such as in a gravel-dominated aquifer), as well as in fractured rock, cave systems, and sinkholes.

Its presence in groundwater indicates that surface water is gaining access to the aquifer, either naturally or induced through pumping. Cryptosporidium may cause serious intestinal illness.

Pathogen survival depends on a number of factors, including soil moisture content (longer survival when moist), soil temperature (longer survival at low temperatures), soil acidity (shorter survival in acid soils), amount of nutrients (longer survival of bacterial pathogen if high), and the activity of microbes that normally live in the soil.

Normal soil bacteria are fully adapted to their environment and should easily outcompete bacterial pathogens for available nutrients. Some will also release one or more chemicals (toxins) that may kill the pathogens.

In most cases, only a small portion of the original pathogen density will reach a well intake unless a rapid pathway exists (such as fractured rock). Despite this, most reported waterborne disease outbreaks reported in the United States each year are associated with groundwater. Between 1971 and 2000, almost 60 per cent of the approximately 700 reported outbreaks, with more than 90,000 cases of illness, were associated with water systems using groundwater.

PROTECTING PUBLIC HEALTH

How can the public be protected from fecally contaminated groundwaters? Drinking-water supplies using groundwater can be protected from fecal contamination of the groundwater in several ways. Most importantly, suppliers can ensure that their well meets the state well-construction code.

They can identify any sources of fecal contamination near the well intake and take proper corrective measures. Suppliers can also have an expert conduct a periodic on-site inspection, including an assessment of how vulnerable the source water is to nearby sources of fecal contamination.

Another means by which a water supply can be protected is to periodically test its source-water quality. To test for all possible pathogens, however, is not practical; the wide variety of potential waterborne pathogens and the limitations of analytical methods would make testing for all such pathogens extremely difficult, time-consuming, and expensive. Instead, inexpensive and easily measured indicators of fecal contamination are used rather than testing for specific pathogens. These include microbes that are extremely common in the gut of humans and other warm-blooded animals, but not elsewhere. If such indicators are absent, the pathogens whose normal habitat is the intestinal tract of humans and other warm-blooded animals should also be absent. Among the fecal indicators used for environmental waters, including groundwater, are Escherichia coli (a bacterium), fecal streptococci (a small group of bacteria), enterococci (very similar to fecal streptococci), and coliphage (a group of viruses that infect and kill E. coli). Water disinfection provides another barrier to pathogen entry and survival in the pipe network that distributes water (distribution system), especially if the source water is contaminated or vulnerable to fecal contamination. Disinfection kills pathogens and other microbes. Commonly used disinfectants for groundwaters include chlorine and ultraviolet light.

The U.S. Environmental Protection Agency develops regulations that apply to every water supplier that serves at least 25 people for at least 60 days during a year (defined as a public water system). One regulation that protects groundwater systems from pathogens is the Total Coliform Rule (TCR). The TCR requires every public water system to monitor its tap water Escherichia coli is a bacteria found in the intestinal tract of warm-blooded animals. Its presence in groundwater indicates that a pathway exists between the land surface and the aquifer, and may indicate that the water is not safe to drink. For a group of closely related bacteria, called total coliforms, at a frequency that depends on the number of people a system serves. While total coliforms are not a fecal indicator, they are used to detect problems in water treatment and in the distribution system. If any sample is total coliform positive, the system must then test for a fecal indicator (normally E. coli).

Another regulation, the Surface Water Treatment Rule (SWTR), in addition to surface waters, also covers groundwater systems that are directly under the influence of surface waters. Among other requirements, the SWTR requires all such systems to disinfect. An important element in both the TCR and the SWTR, as amended, is a requirement for systems to have a periodic on-site inspection (sanitary survey) by a state official or a state-approved agent.

As of 2002, the Environmental Protection Agency was developing another regulation, the Groundwater Rule, which will require groundwater systems to determine whether the source water is fecally contaminated or is vulnerable to such contamination. If so, the system will be required to take corrective measures.

DRINKING WATER TREATMENT

Public water systems vary in the treatment of water that is delivered to the consumer, depending on whether the system uses a groundwater source or a surface-water source. Some public water systems, primarily those that utilize groundwater wells, deliver untreated water directly to the customer's tap. Groundwater is usually less susceptible to contamination than surface water, and therefore requires little or no treatment.

Many large public water systems, however, draw their water from surface sources such as lakes, rivers, and streams, which are vulnerable to many types of contamination. Hence, a multiple-barrier approach to water treatment is most effective in producing drinking water that is free of health risks, meets regulatory requirements, and is palatable (acceptable in taste) to the consumer.

Groundwater

Although groundwater generally is less likely to be contaminated than surface water, it nonetheless may require contaminant removal. Possible groundwater contaminants include naturally occurring inorganic chemicals such as arsenic, fluoride, and nitrate, as well as human-made chemicals (from solvents, fuels and pesticides, for example) that have found their way into the aquifer. Other contaminants that present aesthetic concerns are calcium and magnesium (which produce hardness); iron and manganese (which cause staining of laundry and plumbing fixtures); and hydrogen sulfide gas (which produces the unforgettable "rotten-egg" smell). The word "contaminant" includes any substance in water, including those that are not harmful, even though popular usage often refers only to harmful pollutants.

Hard water is usually treated with the water "softening" method, which utilizes a resin-coated media that exchanges its ion (usually sodium) for the calcium and magnesium ions that are responsible for hardness. The watersoftening media must be regenerated periodically with a brine (very salty) solution.

Even if the water is not hard, inorganic chemicals such as iron and manganese can cause problems that require treatment. Iron and manganese can be oxidized, or brought out of solution, by chemicals such as chlorine.

Settling tanks at a water treatment plant allow coagulated particles to settle to the bottom of the tank. The settled material is periodically drained and collected for eventual disposal in a landfill or potassium permanganate. The oxidized metals can then be filtered out as the water passes through a sand media. A zeolite sand, commonly known as manganese greensand, is usually used for this filtration.

Some groundwaters have taste and odour problems that require treatment in order for the water to be palatable to the user. Chlorine and ozone are effective in reducing hydrogen sulfide odors, and simple aeration can rid the water of some volatile organic compounds, such as solvents or petroleum byproducts.

SURFACE WATER

Many large cities are served by surface-water sources; most of these surface waters are exposed to a variety of contaminants. Common contaminant sources are untreated sewage and run-off from fertilized fields, parking lots, or unprotected watersheds. The primary contaminants of concern are microbes, including Giardia and Cryptosporidium. Effective surface water treatment therefore relies on the multiple-barrier approach, involving a sequence of processes from among the following options.

Pre-Treatment

Simple screens over intake pipes in lakes or rivers can prevent large debris, such as leaves, sticks, or small fish from entering the treatment plant. Waters that exhibit seasonal changes in turbidity or algae growth may require an oxidant such as chlorine, potassium permanganate, or ozone for pre-treatment.

Chemical Feed and Rapid Mix.

Coagulants such as alum, ferric chloride, or synthetic polymers are added to the raw water and mixed rapidly. Chemicals to adjust the pH may also be added at this point. The effectiveness of further treatment depends partly on raw water chemical characteristics such as pH, temperature, and alkalinity.

Coagulation, Flocculation, and Sedimentation

The coagulant chemicals, by the process of coagulation, neutralize the electrical charges of the sediment particles, allowing the binding together of small particles into larger particles that can then be settled or filtered. This process is known as flocculation. Small filters made of pleated paper, carbon, or synthetic resins often are used by consumers. Pleated paper filters can remove sand, dirt, and some iron. Carbon filters are effective against chlorine and organic chemicals such as herbicides and pesticides, and can also remove objectionable tastes and odors.

Some manufacturers of resin-type filters claim a high removal of lead or other inorganic chemicals. The sedimentation process then allows the larger "floc" particles to settle in a basin. The cleaner water at the top of the sedimentation basin flows onto the filter.

Filtration

The filter media is usually a mixture of carefully graded and specified gravels and sands, specifically designed for particle retention. These removed particles include turbidity and biological contaminants such as Giardia and Cryptosporidium. Inorganic chemicals such as arsenic and mercury are also removed by proper coagulation and filtration.

Disinfection

This process is designed to kill disease-causing microorganisms such as bacteria, viruses, and protozoa. Chlorine is most commonly used for this purpose, although ozone and ultraviolet light are increasingly being used in public water systems. The single-celled protozoa (Giardia and Cryptosporidium), which can cause severe intestinal illness, are more resistant to traditional disinfectants than bacteria or viruses, and the disinfection process must therefore be closely monitored.

OTHER TREATMENT TECHNOLOGIES

A number of other treatment technologies are employed both by the public and private sector, depending on the needs and desires of the water users. Granular-activated carbon is used for the elimination of objectionable tastes and odors, as well as for removal of organic chemicals such as pesticides and herbicides.

Sodium carbonate, commonly known as soda ash, can be used to raise the pH of the water to make it less corrosive to pipes and plumbing fixtures. Some public water systems add small amounts of fluoride in order to aid in the prevention of dental cavities in children.

In addition, many small water treatment units, designed to fit under the kitchen sink or in any other convenient location, are used by individual homeowners regardless of the water source. Cartridge filters, membrane filters, and reverse osmosis units are all able to filter out most contaminants that cause health risks and aesthetic concerns. Water softeners are often necessary to treat hard water. Small ozone or ultraviolet-light disinfection units are becoming popular with homeowners who depend on private water supplies of questionable quality. Many different types of water treatment units are designed for the individual homeowner. When considering the purchase of such a unit, the homeowner should have the water tested to determine exactly what type of contaminants are in the water, and what, if anything, needs to be removed.

No one filter works for all contaminants. A filter that removes iron and manganese, for example, will probably not remove herbicides or pesticides. It is also important to remember that all treatment units require regular maintenance to keep them functioning properly and to prevent the buildup and possible breakthrough of the unwanted contaminants.

Ecology, Fresh-Water

Streams, lakes, and wetlands differ profoundly from one another in the conditions they provide as habitats for biological communities. Fundamental characteristics of standing water (a lentic system) or flowing water (a lotic system), the dynamics of its interaction with adjacent land and vegetation, and seasonal fluctuations in water conditions determine characteristic biological assemblages.

STREAMS

The most apparent feature of streams is flowing water. The geomorphology and topography through which streams flow, their channel steepness, the variability in stream-bottom substrates, and the availability of woody debris are among the natural controls that influence flow patterns. The diversity in physical structure and flow among riffles, pools, and glides is reflected in equally diverse aquatic communities.

A flowing-water habitat is known as a lotic ecosystem. A lentic system, such as a lake, is predominantly non-flowing.

In flat valley floodplains, streams naturally meander, or flow in a winding path. In these systems, flooding is an important natural process: nutrient-rich sediments are deposited onto the floodplain, and young fish are protected in quieter areas away from high flow. Terrestrial (land-based) invertebrates become important sources of food during these times. Under the stream and beneath floodplain soils, water flows in the subsurface in the hyporheic zone. Waters are cooled, nutrients are exchanged, and invertebrates are specially adapted to live in this subterranean habitat. A river basin consists of its entire upstream network: its watershed is the landscape, including the stream network and the land it traverses. Riparian (streamside) zones make up the transition between stream and terrestrial environment, encompassing the area where vegetation and streams interact. Riparian plants provide shade, cool the stream, retain soils, and filter nutrients. Fallen leaves decompose and provide nutrients for stream microorganisms and invertebrates.

Changing habitats along the longitudinal stream gradient, from its headwaters to downstream reaches, can be thought of as a stream continuum. Energy possessed by the stream and aquatic communities tend to follow this continuum in predictable ways. Leafy inputs are most important in headwaters, whereas algae are more important as a food base in wider, more sunlit reaches. Nutrients released from streambanks and within the stream are taken up by algae in a downstream spiral of repeated uptake and release.

Macroinvertebrates in the stream continuum are classified into functional feeding groups, depending on how they obtain food. In headwaters, shredding invertebrates, dependent on decomposing leaves, are common. As streams open up, scraping macroinvertebrates that scoop algae off rocks are more abundant. Collector invertebrates either gather or filter out small edible particles everywhere, but in wide, deep rivers they dominate with plankton. As adults many stream invertebrates move into the terrestrial environment.

In stream food webs, primary production (the foundation of food chains) comes from riparian inputs and algae, and most invertebrates are herbivores or detritivores. Top predators are primarily fish but also include birds, amphibians, and humans. Invertebrate predators are relatively few. Fish move easily between habitats, and many feed on both stream and riparian invertebrates. A few birds spear large invertebrates that avoid fish predation. Other birds and bats capture adult aquatic insects that emerge from the stream.

LAKES

Lakes are characterized by basin shape, volume, and depth. Shallow edges where rooted vegetation persists are nursery grounds for fish and diverse invertebrates. The open-water (pelagic) zone is classified into layers, depending on the degree of mixing that occurs. Water changes density with temperature, causing lakes to become layered, or stratified, into temperature zones.

Consequently, significant energy is needed to mix a lake's thermal layers. Deep lakes are stratified into a warmer, upper layer that mixes readily (epilimnion); a middle layer of quickly changing temperatures (metalimnion); and a deep, cool bottom layer that mixes infrequently during the year (hypolimnion).

Wind easily mixes shallow lakes, so these layers either do not persist or do not develop. Phytoplankton, predominantly algae, form the base of a lake's food chain. These primary producers fall into five major categories. These major categories also are represented in streams.

COMMON PHYTOPLANKTON

Group Characteristics

Diatoms (Bacillariophyta) Primary pigments are chlorophyll and carotenes so they appear yellow- brown in color, have cell walls made from silica, often have early season dominance, cell walls may persist for thousands of years in sediments where they are useful for determining historic water characteristics.

Green Algae (Chlorophyta) Primary pigment is chlorophyll so they appear bright green, population growth often follows spring diatoms, are important food for zooplankton, occasionally cause nuisance blooms especially when nitrogen concentrations in the lake are high.

Golden Algae (Chrysophyta) Primary pigments are chlorophyll, carotene and xanthophylls, appear yellow-brown in color, optimize their growth during the relatively cool waters of late winter/early spring, taste and odor problems are often associated with golden algae, may supplement photosynthesis by capturing food particles in the water.

Blue-Green Algae (Cyanobacteria) Primary pigments are chlorophyll, carotene, phycobilins, and xanthophylls, appear blue-green, no chloroplasts-color is distributed evenly throughout cell, are similar biologically to bacteria, often form filamentous or spherical colonies, can form nuisance blooms, often dominate summer plankton of productive lakes, most can regulate their buoyancy, many are nitrogen- fixers, some produce potent toxins.

Dinoflagellates (Dinophyta) Appear dark green to brown to almost black in color, are free- swimming, have two flagella, have forward-spiraling swimming motion, blooms are associated with organic pollution, may cause taste and odor problems in water, marine dinoflagellates are responsible for "red tides."

A community of microscopic zooplankton lives in a lake's upper layer, feeding on single-celled or small colonial algae. In clear, relatively unproductive lakes, zooplankton consume much of the algae. Very productive lakes are much less clear, with abundant algae in blooms of filamentous forms.

Water chemistry plays an important role in lake dynamics, as nutrients influence algal productivity and higher trophic levels. In deep lakes, the bottom layer has little oxygen when it is not mixing, and few organisms survive there. Similarly, very salty lakes contain only a few highly specialized zooplankton.

When fish occur in lakes, they are the dominant predators. Fish are visual predators and select prey on the basis of size. They consume the largest zooplankton, and remaining zooplankton will be small. A trophic cascade occurs as the top predators (fish) depress the next trophic level (zooplankton), releasing high production in a third level (algae). In fishless lakes, invertebrate predators dominate, and large herbivorous zooplankton escape predation.

WETLANDS

Wetlands generally are more shallow than lakes and have standing or flowing water above, near, or below the surface. They vary in size, from puddles to thousands of hectares (a hectare is 10,000 square meters). Because water is not necessarily present year-round, characteristic waterlogged soils or plants help identify their location and extent. Wetlands often are intermediate habitats, representing a transition between aquatic and terrestrial habitats. Their transitional position makes them key exporters of organic materials and sinks for nutrients. Persistent, shallow standing water forms a permanent wetland. When soils are saturated for only part of the year, temporary or ephemeral wetlands are created.

Wetland communities reflect the duration and depth of standing water. Plants vary specifically in requirements for annual wetness. Similarly, invertebrates differ in intervals required to mature, eventually migrating or becoming dormant in the dry season. Some amphibians use only a seasonal aquatic habitat before migrating; others require continuous wetlands.

Ephemeral plant and invertebrate resources will attract migratory waterfowl, but resident birds and fish require year-round water. Thus, a variety of wetlands support diverse plant and animal communities.

Human Influences

Habitat loss due to human activities has severely affected fresh-water ecosystems. Because of filling and draining activities more than half (53 per cent) of wetlands in the continental United States were lost between 1780 and 1980. Dams, channelization, wood removal, and flow alteration have reduced stream habitat significantly. Profound changes from accidental and intentional introduction of non-native plant and animal species have altered aquatic communities and destroyed habitats. Acid rain and other consequences of air and water pollution have severely damaged lakes, streams, and wetlands.

Cumulative effects of these activities often result in declining native flora and fauna. Although scientists and decisionmakers recognize many alterations caused by humans, they have just begun to develop scientific and political understandings for fresh-water restoration.

ALGAL BLOOMS IN FRESH WATER

Aquatic ecologists are concerned with blooms (very high cell densities) of algae in reservoirs, lakes, and streams because their occurrence can have ecological, aesthetic, and human health impacts.

In waterbodies used for water supply, algal blooms can cause physical problems (e.g., clogging screens) or can cause taste and odour problems in waters used for drinking. Blooms involving toxin-producing species can pose serious threats to animals and humans.

Algae in Aquatic Ecosystems

The term "algae" is generally used to refer to a wide variety of different and dissimilar photosynthetic organisms, generally microscopic. Depending on the species, algae can inhabit fresh or salt water.

In modern taxonomic systems, algae are usually assigned to one of six divisions. The misnamed blue-green algae are often grouped with algae because of the chloroplasts contained within the cells. However, these organisms are actually photosynthetic bacteria assigned to the group cyanobacteria.

Fresh-water algae, also called phytoplankton, very in shape and colour, and are found in a large range of habitats, such as ponds, lakes, reservoirs, and streams. They are a natural and essential part of the ecosystem. In these habitats, the phytoplankton are the base of the aquatic food chain. Small fresh- water crustaceans and other small animals consume the phytoplankton and in turn are consumed by larger animals.

Bloom Occurrences and Impact

Under certain conditions, several species of true algae as well as the cyanobacteria are capable of causing various nuisance effects in fresh water, such as excessive accumulations of foams, scums, and discolouration of the water. When the numbers of algae in a lake or a river increase explosively, an algal "bloom" is the result. Lakes, ponds, and slow-moving rivers are most susceptible to blooms.

Algal blooms are natural occurrences, and may occur with regularity (e.g., every summer), depending on weather and water conditions. The likelihood of a bloom depends on local conditions and characteristics of the particular body of water. Blooms generally occur where there are high levels of nutrients present, together with the occurrence of warm, sunny, calm conditions.

However, human activity often can trigger or accelerate algal blooms. Natural sources of nutrients such as phosphorus or nitrogen compounds can be supplemented by a variety of human activities. For example, in rural areas, agricultural run-off from fields can wash fertilizers into the water. In urban areas, nutrient sources can include treated wastewaters from septic systems and sewage treatment plants, and urban stormwater run-off that carries non- point-source pollutants such as lawn fertilizers.

An algal bloom contributes to the natural "aging" process of a lake, and in some lakes can provide important benefits by boosting primary productivity. But in other cases, recurrent or severe blooms can cause dissolved oxygen depletion as the large numbers of dead algae decay.

In highly eutrophic (enriched) lakes, algal blooms may lead to anoxia and fish kills during the summer. In terms of human values, the odors and unattractive appearance of algal blooms can detract from the recreational value of reservoirs, lakes, and streams. Repeated blooms may cause property values of lakeside or riverside tracts to decline.

Toxic Blooms

Some algae produce toxic chemicals that pose a threat to fish, other aquatic organisms, wild and domestic animals, and humans. The toxins are released into the water when the algae die and decay.

The most common and visible nuisance algae in fresh water, and the species that are often toxic, are the cyanobacteria. A cyanobacterial bloom will form on the surface and can accumulate downwind, forming a thick scum that sometimes resembles paint floating on the water. Because these mats are blown close to shore, humans and wild and domestic animals can come into contact with the unsightly material.

Blooms of toxic species of algae and cyanobacteria can flood the water environment with the biotoxin they produce. When toxic, blooms can cause human illnesses such as gastroenteritis (if the toxin is ingested) and lung irritations (if the toxin becomes aerosolized and hence airborne). Other cyanobacterial toxins are less drastic, and cause skin irritation to people who swim through an algal bloom. Toxicity can sometimes cause severe illness and death to animals that consume the biotoxin-containing water.

Cyanobacterial toxins are known to affect bean photosynthesis when they are present in irrigation water. The toxins also can modify zooplankton communities, reduce growth of trout, and interfere with development of fish and amphibians. In some cases, toxins can be bioconcentrated by fresh-water clams.

Some algal blooms in fresh water may only be a nuisance, but others can deplete dissolved oxygen in the water or generate biotoxins that are harmful to birds, fish, and other animals. This Canada goose swims among a floating layer of heavy, but probably harmless, algal growth.

Microcystins comprise the most common group of about fifty cyanobacterial toxins. Among these toxins are ones that, if ingested in sufficient quantity, can harm the liver (hepatotoxins) or nervous system (neurotoxins). Microcystins can persist in water because they are stable in both hot and cold water.

Even boiling the water, which makes the water safe from harmful bacteria, will not destroy microcystins. As a result of this threat, the Canadian government implemented a recommended water-quality guideline of 1.5 ìg per litre of microsystin-LR (the most common hepatotoxin), and other countries will likely follow suit.

In Canada as well as the United States, there are few reports of injury and no reports of human deaths resulting from microcystins in drinking water, in large part because surface-water sources of drinking water (e.g., reservoirs, lakes, and rivers) must undergo filtration and chlorination at water utilities prior to being distributed to customers. (Cyanobacterial toxins can be removed from water only by activated charcoal filters and chlorination.)

Control Considerations

Repeated episodes of algal blooms can be an indication that a river or lake is being contaminated, or that other aspects of a lake's ecology are out of balance. While cyanobacterial blooms receive the most public and scientific attention, the excessive growth of other algae and other aquatic plants also can cause significant degradation of a lake or pond, particularly in waters receiving sewage or agricultural run-off.

Aquatic biologists and other water-quality specialists often are called to identify the causes and recommend management steps to reduce or control the problem.

However, prevention of a problem is always better than trying to fix the problem after it happens. Controlling agricultural, urban, and stormwater run- off; properly maintaining septic systems; and properly managing residential applications of fertilizers are probably the most effective measures that can be taken to help prevent human-induced fresh-water algal blooms.

BIODIVERSITY

Biodiversity describes the variety of biological organisms in a given habitat, area, or ecosystem. It includes several components involving variation in species, ecosystems, and genetics. Species diversity is perhaps what most people think of when they think of biodiversity. Species biodiversity describes the number of different species of plants, animals, fungi, protists, or bacteria within an area of interest. Although there are approximately 1.5 million named species globally, the large majority of species have yet to be studied or named, and there are likely to be as many as 10 million species on Earth.

Ecosystem diversity describes the variety of habitat types found within a given area. For example, there are numerous types of terrestrial (land-based) ecosystems in the United States, including forests, grasslands, wetlands, and deserts. There also are many types of aquatic (water-based) ecosystems, including lakes, rivers, estuaries, coral reefs, and intertidal habitats.

Genetic diversity describes genetic variation within species and represents a third component of biodiversity. Genetic variation allows species to adapt to changes in their environment. Species that are reduced to very small populations lose much of the gene pool and hence can lose much of this adaptability.

Aquatic Biodiversity

Aquatic biodiversity describes the diversity of species and ecosystems found in and around aquatic habitats such as rivers, lakes, and oceans. As with terrestrial ecosystems, aquatic biodiversity varies from region to region. Aquatic biodiversity is greatest in tropical latitudes. For example, an estimated 3,000 species of fish are found in the Amazon River alone.

Coral reef habitats also have extremely high biodiversity; nearly a quarter of all known marine species are found in coral reefs. The Great Barrier Reef, off the coast of Australia, is the largest coral reef system in the world. It supports over 700 species of coral, in addition to 1,600 fish species and 4,000 species of mollusks.

In the Antarctic Ocean, on the other hand, only 120 fish species are found. These species possess special molecular, biochemical "antifreeze" properties to deal with the cold water temperatures. However, Antarctic habitats nonetheless support many unique aquatic groups, such as the albatross, penguin, and large numbers of marine mammals such as the whale and seal.

Many fresh-water habitats also harbour a high proportion of unique species. This is due to the fact that, unlike oceans, fresh-water habitats often are isolated from one another, with natural barriers between them that are the concept of biodiversity can be embodied in this colourful reef fish: a genetically distinct species is found in a specific habitat within a larger ecosystem. This particular fish could not survive, for example, in a polar sea or a tidal river. Difficult to cross. This results in the evolution of distinct species in different fresh-water habitats. The preservation of fresh-water habitats therefore is particularly critical to conserving aquatic biodiversity.

The Value of Biodiversity

The value of biodiversity is an issue that has caused considerable debate, given that the preservation of habitats often conflicts with the desires of developers. Yet there are several reasons for valuing biodiversity.

First, biodiversity is essential to the functioning of ecosystems. Each species plays a unique role within an ecosystem, and every species is dependent on others for food, shelter, or other resources. The loss of a single species therefore can have profound effects for the ecosystem as a whole.

Second, all species are potential sources of genetic variation for the development of new types of agricultural crops, as well as of medical drugs for treatment of human diseases. Third, biota (living organisms) have scientific and educational value. Finally, species have aesthetic and recreational value— consider, for example, the popularity of activities such as snorkeling, scuba diving, and hiking. Scientists have shown that habitats with greater biodiversity are more resilient—that is, they are better able to adjust to and recover from various disturbances. Because different species may perform overlapping functions in a biologically diverse ecosystem, a disturbance that affects one species may have lesser impact on the ecosystem as a whole. Habitats with little diversity are more vulnerable, because a disturbance affecting one species may cause the entire network of interactions to collapse.

Ecosystems approaches to natural resource management address interactions among species and among food webs, as well as the cycling of resources such as carbon, water, and nitrogen. These ecosystems approaches focus not on single species, but on the preservation of complex sets of interactions among species. Preservation of large, intact areas of habitat is necessary for the continued functioning of ecosystems.

Coral reefs are known for their biodiversity as well as their scientific, educational, and recreational value. This brain coral gets its name from its resemblance to the human brain. Even a few centimeters of water can create a small habitat: in this case, a shallow coastal splash-zone pool on a rocky shore. Like tidal pools, such coastal environments support diverse communities of plants and animals.

Threats to Biodiversity

Aquatic biodiversity is threatened on many fronts. Fresh-water habitats support many of the most highly threatened animal groups. These include fish, mussels, amphibians, and crustaceans.

Fresh-water habitats are threatened by many factors, including pollution from industry, increased acidification, and agricultural run-off containing residues of fertilizers or pesticides. In addition, the building of dams destroys many river ecosystems. Development can harm aquatic habitats or remove them altogether, as when marshy areas are filled.

Aquatic ecosystems also are particularly fragile because the disturbance of a watershed can affect multiple components downstream, including rivers, lakes, estuaries, and oceans. Perhaps the largest threat to ocean biodiversity is overfishing.

In addition to depleting commercial species of fish, bivalves, and crustaceans, many fishing methods cause the needless deaths of non-commercial fish species as well as numerous reptiles, birds, and marine mammals.

WELLS AND WELL DRILLING

Approximately 97 per cent of the fresh water available in the world is underground. Wells provide groundwater for individual domestic needs, communities, cities, industry, crop irrigation, and agriculture. Some wells tap hot water, or geothermal resources. In other cases, groundwater is used solely for its cooling capabilities. Some wells are dug solely to study water quality or quantity: these are called monitoring wells or observation wells.

Regardless of its purpose, a well is defined as an artificial hole in the land surface created to access a liquid. It normally has a small diameter, typically less than 3 meters (10 feet), and usually measured in centimeters (inches). Wells may be constructed to seek water, oil, or natural gas.

Parts of a Well

The top portion of a well is commonly called the wellhead. The appearance of the wellhead varies depending on its purpose, when and how it was constructed, and what materials were available when it was built.

A hand-dug well may look like the nursery rhyme well in Jack and Jill— a deep hole surrounded by a stone wall. Most wells however, appear as a pipe, usually 5 to 25 centimeters (2 to 10 inches) in diameter, sticking up a short distance above the ground, typically less than 60 centimeters (2 feet).

In cold climates, the wellhead may be covered by a wellhouse or pumphouse to help minimize potential damage from freezing weather. When a well is constructed in an area where an above-ground structure would create a problem, such as in a road, the wellhead may be in a vault. The vault usually has a strong cement floor, with a drain, cement walls, and a steel top.

The typical components of a well include the sanitary seal, casing, casing seal, well screen, and the pump.

Casing, Screen, and Seals

The sanitary seal is the top covering of a well. Its function is to keep insects, small animals, dirt and debris from entering the well and contaminating the aquifer. This cap usually has an access port that allows periodic water-level measurements, and may have a screened breathing tube to allow air pressure to equalize as water levels rise and fall in the well.

The casing has several important functions. It is a conduit used to hold the wellhole open in soft or unconsolidated geologic formations. It commonly serves as the framework needed to support any pumping mechanism. It is an integral part of the casing seal, which is constructed to prevent surface water and shallow groundwater from entering the well and degrading deeper waters. To prevent surface run-off from flowing into the well, the casing usually extends 30 centimeters (1 foot) or more above the land surface. In formations prone to collapsing, such as in sands and gravel, the casing may extend the entire depth of the well. Where the well is constructed in hard formations such as shale or basalt, only the very top portion of the well may be cased.

Even in ideal circumstances, a 5.5-meter-long (18-footlong) long casing generally is considered the minimum.

The casing seal is an impervious barrier, usually of cement grout or clay, placed between the well casing and the native material. It prevents surface water and undesirable groundwater from other aquifers from entering the wellbore or from flowing down the outside of the casing and degrading deeper water.

Where wells penetrate confining layers at greater depths, seals are commonly constructed at least 1.5 meters (5 feet) into a confining layer. Occasionally, seals are also constructed at greater depths between different aquifers. These seals prevent one aquifer from draining into another, potentially resulting in resource loss or degradation.

Screens in the casing are used to filter out sand and gravel while allowing maximum water flow into the well. Their construction varies from highly engineered screens over 15 meters (50 feet) long, to a few holes punched through the casing. The formations encountered in drilling and the intended use of the well indicate the type of screen, if any, that is needed.

Pumps

While some wells may flow at land surface under artesian pressure, most require some type of pumping mechanism to lift the water. Almost all of the common well pumps are centrifugal in design, using spiral impellers to push or pull water up and out of the well. The most common of these is the above- ground centrifugal, where the motor and pump sit at or near the well. These pumps are capable of moving large volumes of water, but they cannot typically lift from depths greater that 8 meters (25 feet).

Wells with deeper static water levels and relatively low discharge requirements, such as domestic wells, often use submersible pumps. These pumps usually are attached directly above the motor and the entire unit is suspended in the well. Wells with deep water levels used for large quantities of water often use turbine pumps. Turbine pumps have the motors sitting on top of the well with a long shaft turning the impellers deep below the surface.

CONSTRUCTING A WELL

Siting

One of the first things to decide is where to place the well. Assuming that an aquifer has been found at depth, the important considerations include maximizing distances to potential contaminants, access for machinery and potential future uses of the property. Septic systems, waste disposal lines, animal feedlots and buried fuel tanks all have the potential to degrade groundwater quality. Wells should always be at least 1.5 meters (5 feet) from any building site, in an area where future access can be preserved when repair, alteration or abandonment become necessary.

Construction

There are many ways to construct a satisfactory well. The intended use and local geology usually dictate which methods will work in a specific area. Some of the common construction methods include digging, driving, and drilling.

A dug well is built by excavating a hole into a water-bearing zone, lowering a casing into the aquifer, and constructing the seal. Dug wells usually a worker reacts as water surges from a valve as a water well is activated in 1999 in Tekoa, West Bank. The well was the first to be owned and operated by the Palestinian Water Authority on the West Bank, and was financed by the U.S.

Agency for International Development. The water was earmarked for the Bethlehem and Hebron areas are inexpensive, but are restricted to the near- surface, which may be insufficient to provide ample water quantity of suitable quality.

A driven well typically consists of steel point attached to a section of perforated pipe or stiff screen, and sections of solid pipe. After digging or auguring an oversized hole to later accommodate the seal, the well constructor drives the assembly into the earth until the screen or perforations penetrate an aquifer; then he pours the seal. Most wells are constructed with well-drilling equipment. The two main types of drilling machines are the cable tool and the rotary drill.

The cable tool uses a heavy string of tools (bits), dropped repeatedly over the same point, to loosen, fracture and pulverize the earth so that it may then be lifted out of the hole with a bailing bucket. The loosened material and pulverized materials are referred to as drill cuttings. In dry formations, water may be added to the hole to hold the cuttings in suspension so they may be more easily lifted. The bailing bucket is a long cylinder, fitted with a one-way valve at the bottom. The bucket is lowered into the hole, then lifted and drained as the hole is advanced.

Rotary drills use hardened drill bits to cut through the earth. Compressed air, water, or special water-based fluid called drilling mud is used to suspend the cuttings and lift them from the hole. This mud is usually forced down the inside of the drillstem, past the drillbit, where it picks up the cuttings, then up the outside of this drillstem to the land surface. The drilling mud also helps hold the hole open and forms a cake or barrier inside the hole so the fluid and cuttings are forced up the hole rather than into the formation.

Sealing

Regardless of which method is used, care must always be exercised in constructing the casing seal. Improperly sealed wells can contaminate water over a wide area. To construct the casing seal, the well constructor creates an oversized hole down to the intended sealing depth. The casing then is inserted, and the annulus (the space between the casing and earth) is filled with the sealing material. Where the formation is prone to caving, another temporary casing may be necessary to hold the hole open while the seal is constructed. If the sealing depth is shallow (less than 18 meters [50 feet]) and there is little standing water in the sealing area, a special type of clay, known as Bentonite, may be used as the sealing material.

Dry Bentonite chips or pellets are slowly poured into the annular space between the casing and the natural formation. The other common sealing material is cement grout, which is similar to concrete without the gravel component. This fine-grained cement is pumped into the bottom of the annulus until it reaches the land surface.

Developing

Once the well is drilled and the physical components have been constructed, the well must be developed. The purpose of developing the well is to maximize its discharge capacity. This usually is done by alternately pumping the well and surging water in the bore to wash silt and sand from the formation and open it to greater flow rates. A properly developed well will produce adequate quantities of sand and mud-free water.

Testing

After completion of construction, a well test is performed to determine the capacity of the well to yield water over time. This can be done with a pump, a bailer bucket, or by blowing compressed air into the well bore. A component of this test is the monitoring of the well's drawdown.

Drawdown is the change in water level in the well due to pumping. A 1- hour test is considered the minimum period necessary to evaluate a well, with 4-hour and 24-hour tests being more reliable. This information is necessary to determine the usefulness of the well for its desired purpose and the size of pump it can support.

Disindfecting

The well construction process will invariably introduce bacteria into the wellbore. Once construction is completed, the casing, wellbore, and the standing water in the well are sterilized, usually with chlorine or hypochlorate solution.

Once the disinfectant has had sufficient opportunity to sterilize the well, the well is purged, and the fresh water usually is tested for biological and chemical purity.

Logging

During construction, development and testing of the well, it is very important that the constructor record every detail about formations encountered, water-bearing zones, construction components, and well yield.

This record is called the water-well report or well log. It is extremely useful if the well ever needs repair or is to be abandoned. It also provides valuable information to the landowner, other well drillers, and resource managers.

Repair and Abandonment

Over time, old and improperly constructed wells will deteriorate. The landowner is responsible for any wells on his or her property. Wells with faulty seals or that are commingling aquifers may deplete or contaminate the resource. Depending on the original construction methods and materials used, a defective well may be repaired by a skilled well constructor. In many cases, it is easier and cheaper to permanently abandon a defective well than to repair it. Permanent abandonment typically involves removing the existing casing, cleaning out the hole, and completely filling it with cement grout. In situations where the casing can not be removed, the casing must be perforated so the cement grout can penetrate into the formations and prevent any vertical movement of water in the wellbore.

Drilling our Own Well

A water well is much more than a hole in the ground. To prevent groundwater contamination, a well constructor must use proper methods and equipment. Licensed and bonded water-well constructors have the equipment, knowledge, and experience required for proper well construction.

For this reason, most states discourage landowners from drilling a well themselves. If a landowner decides to drill, alter, or abandon a well on his own property, he should first consult with the local regulating authorities to become familiar with specific construction requirement for the area.

MICROBES AROUND US

Microorganisms will grow on simple, cheap medium and will often rise to large populations in a matter of 24 hours. It is easy to isolate their genomic material, manipulate it in the test tube and then place it back into the microbe. Due to their large populations it is possible to identify rare events and then, with the use of powerful selective techniques, isolate interesting bacterial cells and study them. These advantages have made it possible to test hypothesis rapidly.

Using microbes scientists have expanded our knowledge about life. Below are a few examples. Microorganisms have been indispensable instruments for unlocking the secrets of life. The molecular basis of heredity and how this is expressed as proteins was described through work on microorganisms. Due to the similarity of life at the molecular level, this understanding has helped us to learn about all organisms, including ourselves. Some prokaryotes are capable of growing under unimaginably harsh conditions and define the extreme limits of where life can exist. Some species have been found growing at near 100 °C in hot springs and well above that temperature near deep-sea ocean vents.

Others make their living at near 0 °C in freshwater lakes that are buried under the ice of Antarctica. The ability of microbes to live under such extreme conditions is forcing scientists to rethink the requirements necessary to support life. Many now believe it is entirely possible that Jupiter's moon Europa may harbour living communities in waters deep below its icy crust. What may the rest of the universe hold? Until recently, while we could study specific types of bacteria, we lacked a cohesive classification system, so that we could not readily predict the properties of one species based on the known properties of others.

Visual appearance, which is the basis for classification of large organisms, simply does not work with many microbes because there are few distinguishing characteristics for comparison between species even under the microscope. However, analysis of their genetic material in the past 20 years has allowed such classification and spawned a revolution in our thinking about the evolution of bacteria and all other species. The emergence of a new system organizing life on Earth into three domains is attributable to this pioneering work with microorganisms. The fruits of basic research on microbes has been used by scientists to understand microbial activity and therefore to shape our modern world.

Human proteins, especially hormones like insulin and human growth factor, are now produced in bacteria using genetic engineering. Our understanding of the immune system was developed using microbes as tools. Microorganisms also play a role in treating disease and keeping people healthy. Many of the drugs available to treat infectious disease originate from bacteria and fungi. One last recent role of microbes in informing us about our world has been the tools they provide for molecular biology. Enzymes purified from bacterial strains are useful as tools to perform many types of analyses. Such analyses allow us to determine the complete genome sequence of almost any organism and manipulate that DNA in useful ways.

We now know the entire sequence of the human genome, with the exception of regions of repetitive DNA, and this will hopefully lead to medical practices and treatments that improve health. We also know the entire genome sequence of many important pathogens. Analysis of this data will eventually lead to an understanding of the function of critical enzymes in these microbes and the development of tailor-made drugs to stop them. The tools of molecular biology will also affect agriculture.

For example, we now know the complete genome sequence of the plant Arabidopsis (a close relative of broccoli and cauliflower). This opens a new avenue to a better understanding of all plants and hopefully improvements in important crops. Microbes have a profound impact on every facet of human life and everything around us. Pathogens harm us, yet other microbes protect us. Some microbes are pivotal in the growth of food crops, but others can kill the plants or spoil the produce.

Bacteria and fungi eliminate the wastes produced in the environment, but also degrade things we would rather preserve. Clearly they effect many things we find important as humans. In the remainder of this chapter we take a look at how scientists came to be interested in microbes and follow a few important developments in the history of microbiology.

MICROBES ARE HELPFUL

Microbes are much more our friends than our enemies. Although some microbes cause health problems such as strep throat, chickenpox and the common cold, most microbes make our lives better such as: Bacillus thuringiensis - a common soil bacterium that is a natural pest-killer in gardens and on crops. Arbuscular mycorrhizas - fungus living in the soil that helps crops take up nutrients from the soil. Accharomyces cerevisiae - baker's yeast that makes bread rise. Escherichia coli - one of many kinds of microbes that live in our digestive system to help we digest our food every day. Streptomyces- bacteria in soil that makes an antibiotic used to treat infections. Pseudomonas putida - one of many microbes that clean wastes from sewage water at water treatment plants. Lactobacillus acidophilus - one of the bacteria that turn milk into yogurt.

There are many other important jobs microbes do. They are used to make medicine. They break down the oil from oil spills. They make about half of the oxygen we breathe. They are the foundation of the food chain that feeds all living things on earth. We've been using microbes for thousands of years to make products we need and enjoy. For example, we can thank fungi for the cheese on our cheeseburger and yeast for our bun. Cheese and bread are two microbe-made foods people have been enjoying since time began.

In pollution control, researchers are using bacteria that eat methane gas to clean up hazardous waste dumps and landfills. These methane-eating bacteria make an enzyme that can break down more than 250 pollutants into harmless cells.

By piping methane into the soil, researchers can increase growth of the bacteria that normally live in the polluted soil. More bacteria means faster pollution break up.

Also, bacteria is being used as one of the tools to clean up oil spills. These bacteria eat the oil, turning it into carbon dioxide and other harmless by- products. Fungi and bacteria produce antibiotics such as penicillin and tetracycline.

These are medicines we use to fight off harmful bacteria that cause sore throats, ear infections, diarrhea and other discomforts. Scientists have changed the genetic material of bacteria and yeasts to turn them into medicine. They inject genes for medicines they want to make into the microbe cells, as if adding new building information to the microbe's cell DNA. The scientists then grow the microbes in huge containers called fermenters where they reproduce into billions, all making new medicines.

Microbes and the Days of Creation

The world of germs and microbes has received much attention in recent years—and for good reason. We frequently hear the term microbe associated with organisms such as Escherichia coli, Salmonella, anthrax bacteria, antibiotic-resistant tuberculosis, MRSA, HIV, malaria, Stachybotrys and other microscopic creatures. But where do microbes fit into the creation account? Were they created along with the rest of the plants and animals in the first week of creation, or were they created later, after the Fall?

Are microbes a result of the Curse? These and other questions are some that a group of professional creation microbiologists have been asking and their answers may surprise us. Ongoing research based on the creation paradigm appears to provide some answers to these puzzling questions.

Although we cannot be dogmatic about the details of microbe origin during Creation Week, it is believed that a reasonable extrapolation from biological data and Scripture can be made about the nature of microbes in a fully mature creation. Past creation scientists such as Leeuwenhoek, Pasteur and Lister, were blessed by God as He revealed to them critical insight into His creation. So, where do these microbes fit into the very good days of creation? Before answering this question, three terms, microbe, germ and symbiosis, need to be defined. These are relatively new or "modern" terms. First, the Bible does not use these specific terms. These terms were not commonly used until the end of the nineteenth century.

The term microbe was first used in 1878 to describe "extremely minute living beings." Before 1878, scientists including Louis Pasteur, used a variety of terms rather loosely to label the very small organisms that had interested them. It was not clear whether microbes belonged to the animal or plant kingdom, or to a completely different one. The term microbe was given by Charles E. Sedillot to describe bacteria Later, it would also be used of eukaryotic cells, including algae, fungi, protozoans and slime molds. Some people refer to viruses as microbes, but others do not because viruses are not cells. Viruses have non-living, as well as living characteristics.

Today, the term germ refers to disease-causing microbes, or pathogens. All germs would have originated after the Fall The Edenic Curse would have profoundly influenced all creation, including viruses, bacteria, fungi and protozoans that would later become pathogens or parasites. The origin of infectious disease is complex and multifaceted. This topic is further explored in the book.

The Genesis of Germs It provides some understanding into the origin of infectious diseases. From a biblical worldview, infectious diseases and patho- genesis are a secondary state in nature. It is not the way the Creator intended for man and nature. Most microbes are beneficial to man and nature. Only about 5–10 per cent of all bacteria are pathogenic. Many microbes live in a mutualistic relationship with plants, animals, humans and other microbes. Mutualism is a type of symbiosis. The term symbiosis is used to describe an intimate association between organisms of different species. Some symbiotic relationships between microbes and plants, animals and humans are essential for life on earth.

For this reason most microbiologists maintain that bacteria, fungi, protists and other microbes have been maligned in the news media. Without our intestinal flora, we would not digest food nor acquire vitamins and minerals very efficiently.

Without fungi, bacteria, algae and protozoans, life on earth could not last. This is because microbes provide essential "services" (e.g., nitrogen fixation, nitrification and denitrification) in nutrient cycles.

Viruses Fit into Creation

The determination of virus origin is uncertain. It may be that viruses have multiple origins. Some may be degenerate parts from cells after the Curse; still others may have their origin during the days of creation. Today, we think of viruses (Latin for "poison") only in the context of disease. However, some viruses (or at least virus-like genes) are involved in a positive function in nature.

Some groups of viruses, like bacteriophages, play a positive role in controlling bacteria in ecosystems and may play a role in diversity. Another group of viruses play a role in turning off the immune system during pregnancy in mammals and humans. This is a group referred to as endogenous retroviruses (ERVs). ERVs are among a kind of repetitious genetic elements called "retrotransposons". Research has shown that the ERV design prohibits the mother's immune system from damaging the child's body.

These retroviruses cannot fully replicate, only expressed in local immune cells (such as macrophages) of the placenta, thereby preventing them from initiating a full-blown immune response. Thus, the mother's immune system remains competent to respond to other infections but is specifically prevented from mounting an immune response to the developing embryo.

So in creation, the selective ability to turn off the immune system for protection would be a "good" design. Other ERVs also play a positive role in animal and human reproduction. However, since the corruption of creation, the corrupted retrovirus, HIV and various leukemia viruses turn off the entire immune system, leaving the body open to devastating infections.

These examples may provide clues to the origin of viruses and how some may have been created during Creation Week by design and how some have been corrupted as a result of the Fall.

Aseptic Technique

Trying to study this mixed population is often difficult and in the tradition of the scientific method; researchers dissect a system and study each piece in isolation. For microorganisms this means separating the organisms and getting them into pure culture. A pure culture is defined as a growth of microorganisms (a culture) that contains one cell type. It is essential in microbiology to be able to obtain and preserve pure cultures.

Over 100 years ago, Robert Koch devised methods to achieve this goal and the methods he developed are essentially still used today. The goals of aseptic technique are two-fold. The first objective is to obtain pure cultures and secondly to prevent cross-contamination. Microorganisms in culture must not escape into the environment and microbes in the environment must not get into the cultures we are studying. It is essential that aseptic technique be understood and practiced correctly.

Contaminated cultures are worthless for diagnosis or for doing research on, because it is unclear what microbe is performing any action that is being observed. Aseptic methods commonly used are flame sterilization, tube transfer, streak plates, spread plates and pour plates. Flame sterilization is an easy method to insure sterile transfer of a culture from a source to a growth medium.

Tube transfer is useful for moving inocula from one tube to another. Mechanical dilution by making streak plates is the preferred method for obtaining a pure culture of a microorganism. Finally, spread plates and pour plates are common methods for enumerating microorganisms and are sometimes useful for obtaining isolated colonies.

Flame Sterilization and Tube Transfer

Flame sterilization is a very quick simple method of killing microorganisms on an inoculating loop or needle. The loop or needle is held inside a flame for a few seconds to bring it to redness and then cooled. Once cool, the loop or needle can be used for various culture manipulations. Make sure that the area that contacts the culture is flamed to redness. Also, be patient and let the loop cool down, this usually takes about 15-30 seconds. Learning this technique is essential to everything else we do in microbiology.

Transfer of culture from agar plates to tubes, or from tube to tube, is a common, simple procedure. It is important to perform these transfers in a consistent and rapid manner. The following protocols have been found effective. To transfer a culture from an agar plate to a broth or agar slant:

- Place the Bunsen burner in front of we and assemble all necessary equipment with in arms reach. Position everything so that we will not burn ourself while trying to inoculate our tubes.
- Label the tube of broth or agar to be inoculated with identifying marks. The culture, the date and our initials for example. Place it in a rack in front of we.
- Holding the inoculating loop handle, flame the entire wire to redness.
- When the wire cools (about 15-30 seconds) remove the lid of the plate with our other hand and obtain an inoculum by removing a small portion of the surface growth on the agar plate. In most cases we will be picking an isolated colony. Choose a well isolated one. Do not dig into the agar. Replace the lid of the plate immediately.
- Hold the tube to be inoculated with the free hand. Remove the cotton plug or cap of the tube with the little finger of the hand holding the needle holder. If a cotton-plugged tube is used, the mouth of the tube should be passed briefly through the flame to singe off dust and lint particles. (Dust or lint may fall into the tube and contaminate the medium.)
- Introduce the inoculum into the tube. When inoculating a tube of broth, rub the wire against the glass just above the fluid level and then tip the tube slightly to wash the inoculum into the broth. The wire should not be rattled against the sides of the tube to shake an inoculum into the broth; this is unnecessary and may create a dangerous aerosol. If the transfer is made to an agar slant, a single mid-line stroke over its surface is made with the wire or loop.
- Replace the cap or plug.
- Flame the inoculating wire again to redness, slowly to avoid spattering. Put the loop holder down after the wire cools.

Making a Medium

Making medium is as simple as cooking and a crude medium can be made in almost any kitchen with a few utensils and a source of heat. Below is described the production of a chicken broth medium that will grow many common microorganisms.

The medium will be sterilized by tyndaliza-tion. Simply boiling a medium once, while it will kill most vegetative cells, does not kill endospores.

However, heating encourages the endospores to germinate and a second boiling kills these microbes. The boiling process is repeated a third time to ensure that all spores have germinated and been killed.

- Add 250 ml (about 1 cup) of water into a glass container or some other vessel that can stand boiling water. The container should be something we can cover. Glass bottles that can stand boiling or canning jars work well.
- To this add 15 grams, about 1 tablespoon of instant chicken broth crystals. and stir until dissolved. Cover loosely so that steam can escape, but dust and dirt cannot enter.
- Let the broth cool and sit for at least 2 hours up to overnight at room temperature. Make sure that the medium vessel is covered to prevent contamination from the air. At this point the vegetative cells are dead and most endospores will germinate.
- Repeat steps 3 and 4 a second time. This will kill the endospores that germinated on day 1.
- Repeat steps 3 and 4 a third time. This will eliminate the remaining endospores, making the medium sterile.
- Place the medium in a warm place overnight. If our tyndalization was done correctly, our medium should remain clear and free of microbes.
- While we are waiting for our medium, search for a sample we would like to inoculate into it. Almost any sample should work, but natural samples are a good idea for this experiment. Samples coming from food or from our body may contain pathogens. Something we probably want to avoid.
- Check our medium. If it is still clear and sterile, add a pinch of our collected sample to the medium. (The exact amount really does not matter). Again place the medium in a warm place and check it periodically. After a few days, we should start to see the medium become more turbid.
- If we have the equipment, examine a sample of the medium under a microscope. What shapes of microbes are present? Is there more than one species?
- When finished, carefully dispose of the borth. It is basically a spoiled food, but the properties of the spoilage microbe are unknown and should be treated with care.

STREAK PLATES

The streak plate method is a rapid and simple technique of mechanically diluting a relatively large concentration of microorganisms to a small, scattered population of cells. The goal is to obtain isolated colonies on a large part of the agar surface, so that desired species can then be brought into pure culture.

Proper streaking of plates is an indispensable tool in microbiology. In most cases a closed inoculating loop is used for streaking plates. The wire loop should not be badly oxidized or pitted or it will fail to dilute the inoculum and will scratch the surface of the agar.

Streak plates can be made from a broth culture, an agar slant or from an agar plate. It is sometimes convenient to suspend a bit of growth from a solid surface in sterile saline and use this as a source of inoculum. Resuspension of colonies or cultures grown on solid surfaces dilutes the culture and makes streak plating easier.

A loopful of inoculum is transferred from the source and put on the agar surface. When using a large inoculum (a turbid culture or growth from a solid surface), a small spot is spread during the initial transfer. If the inoculum is from a lightly turbid suspension, the first phase of the streaking pattern is begun. The three-phase streaking pattern is recommended for beginners because it is most likely to give satisfactory results with suspensions having a wide range of microbial density.

There are a number of different methods for mechanically diluting microbes on a streak plate. The most common method is spreading microbes across a plate. As the concentration of microbes increases so do the number of phases. Irrespective of the number of phases, loop is flamed between each one. The fifth plate shows an alternative method, where the streaks are not continuous, but are a series of parallel lines. Foobar.

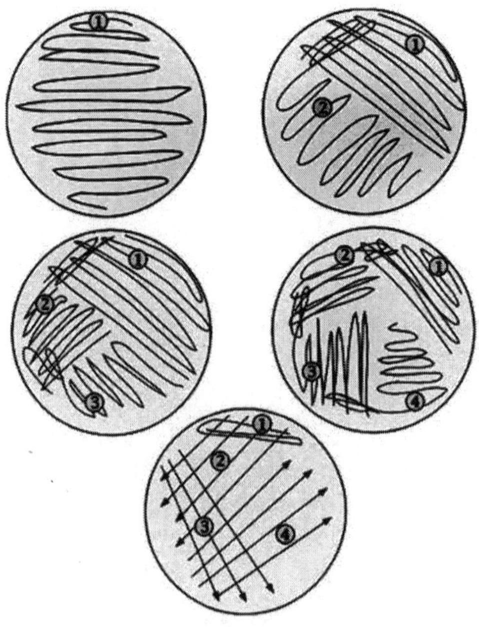

Fig. Streaking Patterns

Choosing a streaking pattern is a matter of individual preference and depends upon the number of microorganisms in the sample. Demonstrates the most common patterns, but they are not the only methods. The object of any streaking pattern is the continuous dilution of the inoculum to give many well isolated colonies. For multi-phase streaking it is crucial to flame the loop before starting the next phase. Note the slight overlap into the previous phase to pick up a small inoculum.

To streak a plate...

- Flame the loop to sterilize it and let cool.
- Position the plate so that the spot of inoculum is nearest the hand not holding the loop (the opposite hand).
- Lift the plate lid with the opposite hand; just enough to get the loop inside and touch the loop to the inoculum spot. It is often helpful to treat the inoculating loop as if it were a pencil - steadying the loop by resting the heel of the hand against the lab bench.
- Move the loop back and forth across the spot and then gradually continue towards the centre of the plate as we sweep back and forth.

Use a very gentle and even pressure.

- When creating each phase, do not worry about keeping each pass across the plate separate from previous ones.
- When about 30 per cent of the plate has been covered by the first streaking phase, remove the loop and flame sterilize it.
- Repeat the above procedure for the second phase, but this time pick up some inoculum by crossing into the first phase 2-3 times and then not passing into it again.
- Repeat as necessary for the third and fourth phases. After streaking the plate, flame sterilize the loop before setting it down.

Spread Plates and Dilution Plating

An absolute requirement for a microbiologist is to be able to determine the concentration of microorganisms in a given sample. Various particle-counting devices, spectrophotometric methods and microscopic techniques have been used to count cells. However, one drawback to these methods is that they count dead as well as living cells.

The most common method of enumerating viable cells is the plate-count method. Diluting microorganisms and placing them into petri plates (or plates) for incubation is another essential technique for working with microorganisms. This method suffers from some problems. First, only those organisms which can grow on the medium and at the temperature and atmospheric conditions of incubation, will divide and develop into colonies. Second, each colony may not represent the progeny from one cell, as two or more cells (those in clusters, chains or otherwise close to one another) can give rise to one colony.

For these reasons the counts obtained from the plate-count method are given as the number of colony-forming units (CFU's) per ml (or gram) rather than the number of cells per ml (or gram). Despite these drawbacks, the plate- count method is a powerful means by which concentrations of viable organisms may be estimated. Also, if it is desirable to count a specific subgroup of microorganisms in a sample, selective media or special incubation conditions can often be used to encourage the growth of only this class of organism.As microbial quantitation involves the use of pipettes (or micropipettes) in preparing dilutions and inoculating plates, the beginning student in microbiology must become familiar with their use.

Due to the possibility of ingesting pathogens and toxic liquids, mouth-pipetting is forbidden in the laboratory! Pipettes are filled and subsequently emptied by the use of propipettes or other pipette bulb. Pay close attention to the demonstration of their use. Important Safety Consideration: When fitting the pipette and pipette bulb together, use very gentle pressure!! Do not jam these items together! (Force is usually not the answer, a good general rule to live by in this lab and in life for that matter.) The glass pipette will probably break and possibly cause severe injury. Handle the pipette only at the top inch or so. For volumes of 5 ml of less, micropipettes are often the tool of choice.

Instruments are available that are capable of dispensing 5 ml all the way to less than 1 µl (one one-millionth of a litre). Micropipettes have made it possible to miniaturize many experiments and greatly decrease the cost of running them. They are also easy to use and can dispense volumes quickly, increasing the number of experiments that can be performed in a set amount of time. Micropipettes are the tool of choice for small volumes.

MICROBIAL ENZYMES

Indians are known to be lovers of milk and its products. As a sequel to white revolution, India has surged ahead to become the largest milk producer in the world, the production figure touched to 74 million tons for the year 1998. With growing urbanization, demand for processed dairy foods has increased considerably, in particular demand for different cheese varieties and low-lactose milk due to increasing intolerance of human beings to lactose in milk and other milk products. For improving the quality of milk and milk products, a number of different enzymes from microbial as well as from non- microbial sources have potential applications in dairy processing.

The use of rennet in cheese manufacture was among the earliest applications of exogenous enzymes in food processing, dating back to approximately 6000 B C. In 1994, the total production of cheese was 8000 metric tons against a total demand of 9000 metric tons. The projected demand by 2000 A.D is around 30,000 metric tons. The use of rennet, as an exogenous enzyme, in cheese manufacture is perhaps the largest single application of enzymes in food processing.

In recent years, proteinases have found additional applications in dairy technology, for example in acceleration of cheese ripening, modification of functional properties and preparation of dietic products. Principal among some enzymes that have important and growing applications are lipases and b- galactosidases. Enzymes with limited applications include glucose oxidase, superoxide dismutase, sulphydryl oxidase, etc.

The increasing use of enzymes to produce specific products with characteristic attributes can be emphasized by the world-wide sale of industrial enzymes approximating to US $ 1.6 billion, which is expected to reach US $ 3.0 billion by the year 2008. Almost 45 per cent of this is shared by the food industry and the remaining is shared by detergent (34.4 per cent), textile (11 per cent), leather (2.8 per cent), pulp and paper (1.2 per cent) industries and other industries (5.6 per cent) excluding enzymes for use in diagnostics and therapeutics. India being the highest producer of milk in the world and consequently the surplus availability of milk in our country has triggered the food and dairy industry to convert the liquid milk into value-added products using biochemical and enzymatic processes.

MICROBIAL RENNETS

Animal rennet (bovine chymosin) is conventionally used as a milk-clotting agent in dairy industry for the manufacture of quality cheeses with good flavour and texture. Owing to an increase in demand for cheese production world wide – i.e. 4 per cent per annum over the past 20 years, approximating 13.533 million tons – coupled with reduced supply of calf rennet, has therefore led to a search for rennet substitutes, such as microbial rennets. At present, microbial rennet is used for one-third of all the cheese produced world wide.

Rennin acts on the milk protein in two stages, by enzymatic and by non- enzymatic action, resulting in coagulation of milk. In the enzymatic phase, the resultant milk becomes a gel due to the influence of calcium ions and the temperature used in the process. Many microorganisms are known to produce rennet-like proteinases which can substitute the calf rennet.

Microorganisms like Rhizomucor pusillus, R. miehei, Endothia parasitica, Aspergillus oryzae and Irpex lactis are used extensively for rennet production in cheese manufacture. Extensive research that has been carried out so far on rennet substitutes has been reviewed by several authors. Different strains of species of Mucor are often used for the production of microbial rennets.

Whereas best yields of the milk-clotting protease from Rhizomucor pusillus are obtained from semisolid cultures containing 50 per cent wheat bran, R. miehei and Endothia parasitica are well suited for submerged cultivation. Using the former, good yields of milk-clotting protease may be obtained in a medium containing 4 per cent potato starch, 3 per cent soybean meal and 10 per cent barley. During growth, lipase is secreted together with the protease. Therefore, the lipase activity has to be destroyed by reducing the pH, before the preparation can be used as cheese rennet.

Microbial rennets from various microorganisms (marketed under the trade names such as Rennilase, Fromase, Marzyme, Hanilase, etc.) being marketed since the 1970s have proved satisfactory for the production of different kinds of cheese. The molecular and enzymatic properties of chymosins have been studied extensively. Although the proteolytic specificities of the three commonly used fungal rennets are considerably different from those of calf chymosin, these rennets have been used to produce acceptable cheeses. Recently Novo Nordisk has succeeded in expressing just one proteolytic enzyme from the fungus R. miehei in the well-known organism A. oryzae. This host organism is able to produce the single protease that cleaves the casein into a glycomacropeptide and para casein by hydrolysing only at the phe105– met106 peptide bond between phenyl alanine and methionine. This monocomponent enzyme product has the trade name Novoren. One major drawback of microbial rennet use in cheese manufacture, is the development of off flavour and bitter taste in the non-ripened as well as in the ripened cheeses.

The rennets from microbial sources are more proteolytic in nature in comparison to rennet from animal sources, resulting in production of some bitter peptides during the process of cheese ripening. Hence, attempts have been made to clone the gene for calf chymosin and to express it in selected bacteria, yeasts and molds.

Recombinant Rennets

Due to shortage of calf stomachs and the economic value of cheese rennet, gene for calf chymosin was one of the first genes for mammalian enzymes that was cloned and expressed in microorganisms. Many different laboratories have cloned the gene for calf prochymosin in Escherichia coli and analysed the structure of the gene as well as the properties of the recombinant chymosin. The expressed proenzyme in E. coli, is present mainly as insoluble inclusion bodies, comprised of reduced prochymosin as well as molecules that are interlinked by disulphide bridges. After disintegration of the cells, inclusion bodies are harvested by centrifugation.

The individual laboratories have reported some differences in the procedure for renaturation of prochymosin from the inclusion bodies, but all have followed the same general scheme. The enzymatic properties of recombinant E. coli chymosin are indistinguishable from those of native calf chymosin. The enzymes were identical when observed by immunodiffusion in gels, but a slight difference was observed by enzyme linked immunosorbent assay (ELISA).

Table. Recombinal Chymosin Proparations at/or Approaching Legal and Commercial Acceptance.

Sourcc of DNA microorganisms	Producing (brand name)	Producing company
Calf abomasums	klayveromyces lactis (Maxiren)	Gist Brocades
Calf abomasums Hansen	Aspergillus niger	Genccor/chr.
Synthetic	(Chyrnoger) Escherichic coli	Pfizer (Chymax)

The gene for prochymosin has also been cloned in Saccharomyces cerevisiae; the levels of expression have been reported to be 0.5 to 2.0 per cent of total yeast protein. In yeast, about 20 per cent of the prochymosin can be released in soluble form which can be activated directly; the remaining 80 per cent is still associated with the cell debris.

The cloning of chymosin was carried out without the prosequence. Though the level of chymosin mRNA was similar to that of prochymosin mRNA, no milk-clotting activity was observed in clones containing the chymosin gene only. The results suggest that the prosequence is essential for correct folding of the polypeptide chain. The zymogen for the aspartic proteases from R. pusillus, also called mucor rennin, has likewise been cloned and expressed in yeast. Studies on its conversion to active form showed that secretion of R. pusillus protease from recombinant yeast was dependent on glycosylation of the enzyme. Compared to yeast, the filamentous fungi generally secrete larger quantities of proteins into the culture medium. Furthermore, filamentous fungi secrete the heterologous proteins with correct folding of the polypeptide chain and process correct pairing of sulphydryl groups.

The gene for R. miehei protease has been expressed in A. oryzae. Prochymosin has been expressed in Kluyveromyces lactis, A. nidulans, A. niger and Trichoderma reesei. In most of the cases, the reported yields of the model systems were about 10–40 mg of enzyme per litre of culture medium. However, 3.3 g of enzyme per litre has also been achieved. Yeast, Kluyveromyces lactis, has recently been used as an efficient host for the secretion of recombinant chymosin, which has led to the development of large-scale production process for chymosin. If produced on an industrial scale, the yields will perhaps be of the latter magnitude.

Chymosins which are approaching or have marketing and legal acceptance and generally regarded as safe (GRAS). Most of the companies produce recombinant rennet of cattle calf origin in different microbial hosts, however in India, the major source of milk is buffalo, showing a different composition from that of cow. The rennet from buffalo source has inherent compatibility for clotting buffalo milk. Therefore, National Dairy Research Institute, Karnal, has taken a lead in cloning gene of buffalo chymosin.

The buffalo chymosin cDNA has been cloned in E. coli and partial N- terminal sequencing of the purified buffalo chymosin indicates that it is highly homologous to the cattle chymosin. Several cheese making experiments have been carried out with recombinant chymosin and the general aspects of recombinant chymosin have been dealt with in a report. Since most of the rennet (> 90 per cent) added to cheese milk is lost in the whey, immobilization would considerably extend its catalytic life. Several rennets have been immobilized, but their efficiency as milk coagulants has been questioned.

So, there is a fairly general support for the view that immobilized enzymes cannot coagulate milk properly owing to inaccessability of the Phe–Met peptide bond of k-casein and that the apparent coagulating activity of immobilized rennets is due to leaching of the enzyme from the support. Different types of conventional cheeses have been successfully made by using recombinant rennet on an experimental or pilot scale.

No major differences have been detected between cheeses made with recombinant chymosins or natural enzymes, regarding cheese yield, texture, smell, taste and ripening. The recombinant chymosins are identical with calf rennet according to the report on biochemical and genetic evidences.

Dairy Industry

Lactose, the sugar found in milk and whey and its corresponding hydrolase, lactase or b -galactosidase, have been extensively researched during the past decade. This is because of the enzyme immobilization technique which has given new and interesting possibilities for the utilization of this sugar. Because of intestinal enzyme insufficiency, some individuals and even a population, show lactose intolerance and difficulty in consuming milk and dairy products.

Hence, low-lactose or lactose-free food aid programme is essential for lactose-intolerant people to prevent severe tissue dehydration, diarrhoea and, at times, even death. Another advantage of lactase-treated milk is the increased sweetness of the resultant milk, thereby avoiding the requirement for addition of sugars in the manufacture of flavoured milk drinks.

Manufacturers of ice cream, yoghurt and frozen desserts use lactase to improve scoop and creaminess, sweetness and digestibility and to reduce sandiness due to crystallization of lactose in concentrated preparations. Cheese manufactured from hydrolysed milk ripens more quickly than the cheese manufactured from normal milk.

Technologically, lactose crystallizes easily which sets limits to certain processes in the dairy industry and the use of lactase to overcome this problem has not reached its fullest potential because of the associated high costs. Moreover, the main problem associated with discharging large quantities of cheese whey is that it pollutes the environment. But, the discharged whey could be exploited as an alternate cheap source of lactose for the production of lactic acid by fermentation. The whey permeate, which is a by-product in the manufacture of whey protein concentrates, by ultrafiltration could be fermented efficiently by Lactobacillus bulgaricus.

Lactose can be obtained from various sources like plants, animal organs, bacteria, yeasts (intracellular enzyme), or molds. Some of these sources are used for commercial enzyme preparations. Lactase preparations from A. niger, A. oryzae and Kluyveromyces lactis are considered safe because these sources already have a history of safe use and have been subjected to numerous safety tests. The most investigated E. coli lactase is not used in food processing because of its cost and toxicity problems.

LACTASE

The properties of the enzyme depend on its source. Temperature and pH optima differ from source to source and with the type of particular commercial preparation. Immobilization of the enzymes, method of immobilization and type of carrier can also influence these optima values. In general, fungal lactase have pH optima in the acidic range 2.5–4.5 and yeast and bacterial lactases in the neutral region 6–7 and 6.5–7.5, respectively. The variation in pH optima of lactases makes them suitable for specific applications, for example fungal lactases are used for acid whey hydrolysis, while yeast and bacterial lactases are suitable for milk (pH 6.6) and sweet whey (pH 6.1) hydrolysis. Product inhibition, e.g. inhibition by galactose, is another property which also depends on the source of lactase. The enzyme from A. niger is more strongly inhibited by galactose than that from A. oryzae. This inhibition can be overcome by hydrolysing lactose at low concentrations by using immobilized enzyme systems or by recovering the enzyme using ultrafiltration after batch hydrolysis. Lactase from Bacillus species are superior with respect to thermostability, pH operation range, product inhibition and sensitivity against high-substrate concentration.

Thermostable enzymes, able to retain their activity at 60°C or above for prolonged periods, have two distinct advantages viz. they give higher conversion rate or shorter residence time for a given conversion rate and the process is less prone to microbial contamination due to higher operating temperature. Bacillus species have a pH optima of 6.8 and a temperature optima of 65°C. Its high activity for skim milk and less inhibition by galactose has made it suitable for use as a production organism for lactase.

The enzymatic hydrolysis of lactose can be achieved either by free enzymes, usually in batch fermentation process, or by immobilized enzymes or even by immobilized whole cells producing intracellular enzyme. Although numerous hydrolysis systems have been investigated, only few of them have been scaled up with success and even fewer have been applied at an industrial or semi-industrial level. Several acid hydrolysis systems have been developed to industry-scale level. Large-scale systems which use free enzyme process have been developed for processing of UHT-milk and processing of whey, using K. lactis lactase (Maxilact, Lactozyme). Several commercial immobilized systems have been developed for commercial exploitation. Snamprogretti process of industrial-scale milk processing technology in Italy is one such working systems. They make use of fibre-entrapped yeast lactase in a batch process and the milk used is previously sterilized by UHT.

For pilot plants, there are three other processes designed and developed to handle milk:

- Gist-Brocades, Rohm GmbH (Germany);
- Sumitomo, Japan. These are continuous processes with short residence times. Processing of whey UF-permeate is accomplished by the system developed by Corning Glass, Connecticut, Lehigh, Valio and Amerace corp. The process by Corning Glass is being applied at commercial scale in the bakers yeast production using hydrolysed whey.

Microbial Enzymes

Cheese ripening is a complex process mediated by biochemical and biophysical changes during which a bland curd is developed into a mature cheese with characteristic flavour, texture and aroma. The desirable attributes are produced by the partial and gradual breakdown of carbohydrates, lipids and proteins during ripening, mediated by several agents, viz.

- Residual coagulants,
- Starter bacteria and their enzymes,
- Non-starter bacteria and their enzymes,
- Indegenous milk enzymes, especially proteinases;
- Secondary inocula with their enzymes.

Proteolysis occurs in all the cheese varieties and is a prerequisite for characteristic flavour development that can be regulated by proper use of the above agents. Cheese ripening is essentially an enzymatic process which can be accelerated by augmenting activity of the key enzymes. This has the advantage of initiating more specific action for flavour development compared to use of elevated temperatures that can result in accelerating undesirable non-specific reactions and consequently off flavour development.

Enzymes may be added to develop specific flavours in cheeses, for example lipase addition for the development of Parmesan or Blue-type cheese flavours. Attempts to accelerate the multiple secondary flavour-forming- reactions, e.g. Strecker degradation, have been scarce. The pathways leading to the formation of flavour compounds are largely unknown and therefore the use of exogenous enzymes to accelerate ripening is mostly an empirical process. Different microbial enzymes used to accelerate cheese ripening.

Studies on the hydrolysis of whole casein and isolated casein components to observe the kinetics and specificity of aspartic proteases in rennin, pepsin and four microbial rennet substitutes indicated the considerable differences in the reaction velocity and the extent of hydrolysis on the rennet curd yield. The rennet enzymes were active even in the later phases of cheese-ripening.

PROTEINASES AND PEPTIDASES

Proteolysis is characteristic of most cheese varieties and is indispensable for good flavour and textural development. Proteinases used in cheese processing include:

- Plasmin,
- Rennet;
- Proteinases (cell wall and/or intracellular) of the starter and non-starter bacteria. Approximately 6 per cent of the rennet added to cheese milk remains in the curd after manufacture and contributes significantly to proteolysis during ripening. Combinations of individual neutral proteinases and microbial peptidases intensified cheese flavour and when used in combinations with microbial rennets reduced the intensity of bitterness caused by the latter. Acid proteases in isolation cause intense bitterness.

Various animal or microbial lipases gave pronounced cheese flavour, low bitterness and strong rancidity, while lipases in combination with proteinases and/or peptidases give good cheese flavour with low levels of bitterness. In a more balanced approach to the acceleration of cheese ripening using mixtures of proteinases and peptidases, attenuated starter cells or cell-free extracts (CFE) are being favoured.

Table. Microbial Enzymes Used to Ucoclcrate Chocse Ripening

Enzyme	Source
Microbial lipases	Aspergillus niger A. oryzae Rhizomucor miehei
Lactasce	Streptococcus lactis
Ktuyveromyces sp.	Escherichia coli
Microbial serline proteinases	A. niger
Neutral pmroteinases	Bacillus subtilis A. oryzae

PROTEOLYTIC ENZYMES

The proteolytic system of lactic acid bacteria is essential for their growth in milk and contributes significantly to flavour development in fermented milk products. Lactic acid bacteria have a complex proteolytic system capable of converting milk casein to the free amino acids and peptides necessary for their growth. These proteinases include extracellular proteinases, endopeptidases, aminopeptidases, tripeptidases and proline-specific peptidases, which are all serine proteases. Apart from lactic streptococcal proteinases, several other proteinases from non-lactostreptococcal origin have been reported. There are also serine type of proteinases, e.g. proteinases from Lactobacillus acidophillus, L. plantarum, L. delbrueckii sp. bulgaricus, L. lactis and L. helveticus. Aminopeptidases are important for the development of flavour in fermented milk products, since they are capable of releasing single amino acid residues from oligopeptides formed by extracellular proteinase activity.

Other Dairy Enzymes

Other enzymes used for dairy food application include:

- Proteases to reduce allergic properties of cow milk products for infants;
- Lipases for development of lipolytic flavours in speciality cheeses.

The functional properties of milk proteins may be improved by limited proteolysis through the enzymatic modification of milk proteins. An acid- soluble casein, free of off flavour and suitable for incorporation into beverages and other acid foods, has been prepared by limited proteolysis. The antigenicity of casein is destroyed by proteolysis and the hydrolysate is suitable for use in milk-protein-based foods for infants allergic to cow milk.

Lipolysis makes an important contribution to swiss cheese flavours, due mainly to the lipolytic enzymes of the starter cultures. The characteristic peppery flavour of Blue cheese is due to short-chain fatty acids and methyl ketones. Most of the lipolysis in Blue cheese is catalysed by Penicillium roqueforti lipase, with a lesser contribution from indigenous milk lipase.

The NOVO process for production of enzyme modified cheese (EMC) uses medium-aged cheese which is emulsified, homogenized and pasteurized, after which 'palatase' (a lipase from R. miehei) is added, with or without a proteinase and the blend is ripened at a high temperature for one to four days. The mixture is reheated, a paste results which is suitable for inclusion in soups, dips, dressings, or snack foods. EMC technology has been developed to produce a range of characteristic cheese flavours and flavour intensities, for example swiss, blue, cheddar, provolo-nemor or romano, suitable for inclusion at low levels in many products.

The global market for the production of microbial enzymes for use in dairy-products manufacture is considerably large, but is being dominated only by a limited number of enzyme producers. In India the microbial dairy enzymes requirement has been very limited till now. However, with the advent of technological processes for the manufacture of different varieties of milk products, such as cheeses by the State Dairy Federations, Co-operatives and Private Dairy Product Manufacturers like Amul, Vijaya, Verka, Dynamix, Nestle, Smith Kline Beechem, etc., the markets for the sale of such products in megacities and towns has been slowly growing for the past two to three years.

Presently, many of these microbial enzymes, such as microbial rennets and other enzymes, are being imported. Hence, there is a scope for the production of enzymes such as microbial rennet, lactase, proteinases and lipases indigenously. In the near future, the requirement for these enzymes is bound to increase by leaps and bounds, basically due to requirement of value- added dairy products in the country.

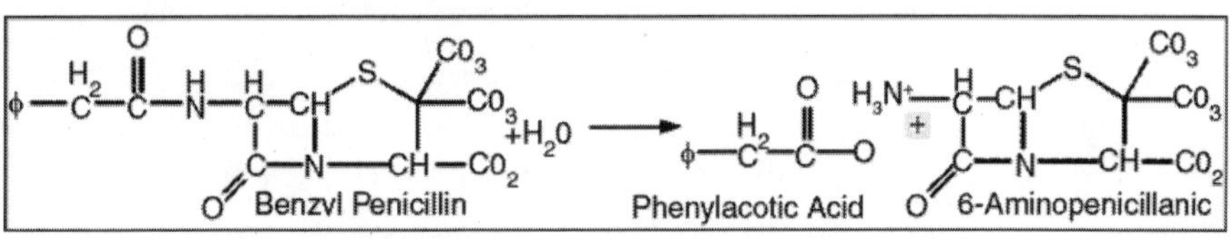

PRODUCTION OF ANTIBIOTICS

The mass production of antibiotics began during World War II with streptomycin and penicillin. Now most antibiotics are produced by staged fermentations in which strains of microorganisms producing high yields are grown under optimum conditions in nutrient media in fermentation tanks holding several thousand gallons.

The mold is strained out of the fermentation broth and then the antibiotic is removed from the broth by filtration, precipitation and other separation methods. In some cases new antibiotics are laboratory synthesized, while many antibiotics are produced by chemically modifying natural substances; many such derivatives are more effective than the natural substances against infecting organisms or are better absorbed by the body, e.g., some semisynthetic penicillins are effective against bacteria resistant to the parent substance.

Benzylpenicillins and phenoxymethylpenicillins (penicillins G and V, respectively) are produced by fermentation and are the basic precursors of a wide range of semi-synthetic antibiotics, e.g. ampicillin. Such specific hydrolysis may be simply achieved by use of penicillin amidases (also called penicillin acylases).

Different enzyme preparations are generally used for the hydrolysis of the penicillins G and V, pencillin-V-amidase being much more specific than pencillin-G-amidase. Penicillin amidase may be obtained from E. coli and has been immobilised on a number of supports including cyanogen bromide- activated Sephadex G200.

It represents one of the earliest successful processes involving immobilised enzymes and is generally used in batch or semicontinuous STR processes (40,000 Ukgpenicillin G, 35°C, pH 7.8, 2 h) where it may be reused over 100 times. It has also been used in PBRs, where it has an active life of over 100 days, producing about two tonnes of 6-aminopenicillanic acid kg-1of immobolised enzyme.

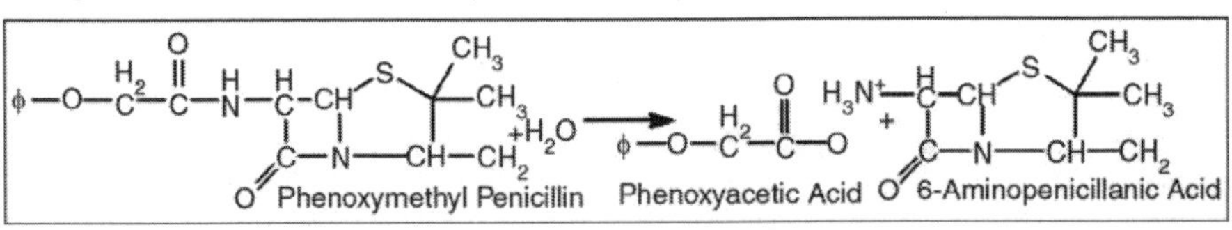

The penicillin-G-amidases may be used 'in reverse' to synthesise penicillin and cephalosporin antibiotics by non -equilibrium kinetically controlled reactions. Ampicillin has been produced by the use of penicillin-G-amidase immobilised by adsorption to DEAE -cellulose in a packed bed column:

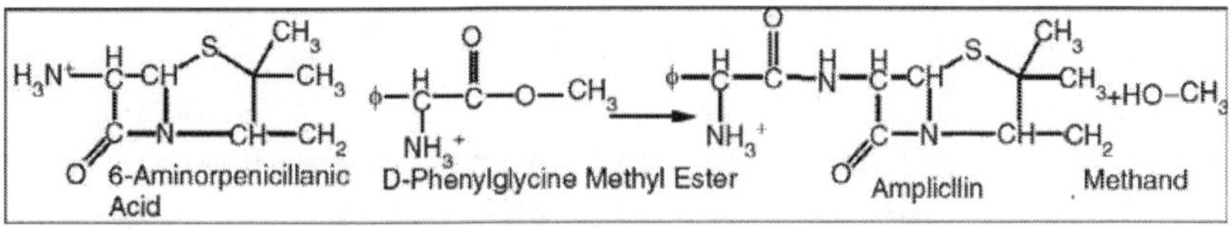

Many other potential and proven antibiotics have been synthesised in this manner, using a variety of synthetic b-lactams and activated carboxylic acids.

PREPARATION OF ACRYLAMIDE

Acrylamide is an important monomer needed for the production of a range of economically useful polymeric materials. It may be produced by the addition of water to acrylonitrile.

$$CH_2{=}CHCN + H_2O \rightarrow CH_2{=}CHCONH_2$$

This process may be achieved by the use of a reduced copper catalyst (Cu+); however, the yield is poor, unwanted polymerisation or conversion to acrylic acid ($CH_2{=}CHCOOH$) may occur at the relatively high temperatures involved (80 -140°C) and the catalyst is difficult to regenerate. These problems may be overcome by the use of immobilised nitrile hydratase (often erroneously called a nitrilase). The enzyme from Rhodococcus has been used by the Nitto Chemical Industry Co. Ltd, as it contains only very low amidase activity which otherwise would produce unwanted acrylic acid from the acrylamide.

Immobilised nitrile hydratase is simply prepared by entrapping the intact cells in a cross-linked 10 per cent (w/v) polyacrylamide/dimethylaminoethylmethacrylate gel and granulating the product. It is used at 10°C and pH 8.0-8.5 in a semibatchwise process, keeping the substrate acrylonitrile concentration below 3 per cent (w/v). Using 1 per cent (w/v) immobilised-enzyme concentration (about 50,000 U l-1) the process takes about a day.

Product concentrations of up to 20 per cent (w/v) acrylamide have been achieved, containing negligible substrate and less than 0.02 per cent (w/w) acrylic acid. Acrylamide production using this method is about 4000 tonnes per year. The closely related enzymes cyanidase and cyanide hydratase are used to remove cyanide from industrial waste and in the detoxification of feeds and foodstuffs containing amygdalin.

$$HCN + 2H_2O \rightarrow HCOO^- + NH_4^+ \quad HCN + H_2O \rightarrow HCONH_2$$

IDENTIFYING USEFUL ANTIBIOTICS

Despite the wide variety of known antibiotics, less than 1 per cent of antimicrobial agents have any medical or commercial value. The most commonly known antibiotic, Penicillin has a highly selective toxicity and therapeutic index (as eukaryotic animal cells do not contain peptidoglycan, they are usually unaffected by it). This is not so for many antibiotics.

Others simply lack advantage over the antibiotics already in use, or have no other practical applications.In order to identify the useful antibiotics, a process of screening is often employed. Using this method, isolates of a large number of microorganisms are cultured and then tested for production of diffusible products which inhibit the growth of test organisms.

However, the majority of the resulting antibiotics are already known and must therefore be disregarded. The remainders must be tested for their selective toxicities and therapeutic activities and the best candidates can be examined and possibly modified.A more modern version of this approach is a rational design programme. This involves screening being directed towards finding new natural products that inhibit specific targets (e.g. a particular step of a metabolic pathway) on microorganisms, rather than tests to show general inhibition of a culture.

Industrial Production Techniques

Antibiotics are produced industrially by a process of fermentation, where the source microorganism is grown in large containers (100,000–150,000 litres or more) containing a liquid growth medium. Oxygen concentration, temperature, pH and nutrient levels must be optimal and are closely monitored and adjusted if necessary. As antibiotics are secondary metabolites, the population size must be controlled very carefully to ensure that maximum yield is obtained before the cells die. Once the process is complete, the antibiotic must be extracted and purified to a crystalline product. This is simpler to achieve if the antibiotic is soluble in organic solvent. Otherwise it must first be removed by ion exchange, adsorption or chemical precipitation.

Strains used for Production

Microorganisms used in fermentation are rarely identical to the wild type. This is because species are often genetically modified to yield the maximum amounts of antibiotics. Mutation is often used and is encouraged by introducing mutagens such as ultraviolet radiation, x-rays or certain chemicals. Selection and further reproduction of the higher yielding strains over many generations can raise yields by 20-fold or more.

Another technique used to increase yields is gene amplification, where copies of genes coding for proteins involved in the antibiotic production can be inserted back into a cell, via vectors such as plasmids. This process must be closely linked with retesting of antibiotic production and effectiveness.

TYPES OF ANTIBIOTICS

The great number of diverse antibiotics currently available can be classified in different ways, e.g., by their chemical structure, their microbial origin, or their mode of action. They are also frequently designated by their effective range.

Tetracyclines, the most widely used broad-spectrum antibiotics, are effective against both Gram-positive and Gram-negative bacteria, as well as against rickettsias and psittacosis-causing organisms. Ciprofloxacin (Cipro) is another broad-spectrum antibiotic, effective in the treatment of mild infections of the urinary tract and sinuses. The medium-spectrum antibiotics bacitracin, the erythromycins, penicillin and the cephalosporins are effective primarily against Gram-positive bacteria, although the streptomycin group is effective against some Gram-negative and Gram-positive bacteria. Polymixins are narrow-spectrum antibiotics effective against only a few species of bacteria.

Administration and Side Effects

Antibiotics are either injected, given orally, or applied to the skin in ointment form. Many, while potent anti-infective agents, also cause toxic side effects. Some, like penicillin, are highly allergenic and can cause skin rashes, shock and other manifestations of allergic sensitivity. Others, such as the tetracyclines, cause major changes in the intestinal bacterial population and can result in superinfection by fungi and other microorganisms. Chloramphenicol, which is now restricted in use, produces severe blood diseases and use of streptomycin can result in ear and kidney damage. Many antibiotics are less effective than formerly because antibiotic-resistant strains of microorganisms have emerged.

Non-medical Use

Antibiotics have found wide non-medical use. Some are used in animal husbandry, along with vitamin B12, to enhance the weight gain of livestock. However, some authorities believe the addition of antibiotics to animal feeds is dangerous because continuous low exposure to the antibiotic can sensitize humans to the drug and make them unable to take the substance later for the treatment of infection. In addition low levels of antibiotics in animal feed encourage the emergence of antibiotic-resistant strains of microorganisms.

Drug resistance has been shown to be carried by a genetic particle transmissible from one strain of micro-organism to another and the presence of low levels of antibiotics can actually cause an increase in the number of such particles in the bacterial population and increase the probability that such particles will be transferred to pathogenic, or disease-causing, strains. Antibiotics have also been used to treat plant diseases such as bacteria-caused infections in tomatoes, potatoes and fruit trees. The substances are also used in experimental research.

❐

Index

A

Abandonment 160, 162
Acid Rain 215
Acinetobacter 188
Aldehydes 41, 243
Anaerobic Protozoans 136
Aquatic Ecosystems 157
Atoms 11, 12, 13, 14
Azotobacter 110

B

Bacilli 19
Biotechnology 57
Bioterrorism 116
Biotreatment 56, 188
Bloodstream 226, 228
Bloom 151, 152, 153

C

Cephalosporin 56, 204, 207
Chemoautotrophic 69
Chlorine 53, 96, 168
Clostridium 47, 49
Cyanobacteria 73, 138, 151, 152, 192
Cytoplasm 13, 15, 36, 41, 111

D

Dalfopristin 100
Degradation 54, 55, 154, 159, 188
Detergent 66
Diversity 147, 154, 156, 188, 192
Dry Heat 65
Duringchemotaxis 21

E

Ecology 77, 147, 153, 192
Enterobacteriaceae 47
Enzymatic Attack 136
Eukaryotic Chemotaxis 62
Evolution 54, 55, 56, 84, 155
Experiment 6, 8, 170
Exposure 65, 66
Extremophiles 192

F

Fermentation 13
Flagellum 135
Fleas 2, 75, 152
Flocculation 145
Flooding 148

G

General Procedures 66
Generation 2, 5, 6, 7, 66
Genetic Diversity 154, 192
Glutamic 14

H

Habitats 93, 94, 140, 147, 148
Halogens 41
Human Health 151
Hydration 114, 115
Hypochlorites 97

I

Immune System 167
Indicators 114, 115
Infection 115
Infectious Disease 4, 166
Insidiousness 88

L

Lactobacillus 192
Legionella 192
Lithotrophic 53

M

Marine 56, 154, 155, 157, 187, 188
Mechanism 61, 62, 63, 158, 159

N

Natural Communities 192
Nucleic 15, 32, 39, 110

O

Observations 2, 3, 5, 7, 8
Optimum 53, 56, 70, 83
Organelles 92, 93
Osteomyelitis 128

P

Pathogen Movement 141
Phospholipid 135
Photosynthesis 10, 13, 93, 110
Public Health 139, 140, 142

Q

Quinupristin 58, 100

R

Receptors 60, 62
Replication 19, 44, 45, 137, 139, 140
Respiration 13, 33, 110

S

Saccharomyces 91, 102
Sedimentation 145, 146
Signal Transduction 60
Spread Plates 172
Surface Water 143, 145

T

Taxonomy 9
Termednecrotizing 129
Toxic Blooms 152

U

Ulcerative Enteritis 149

V

Vaccines 29, 60, 75, 150
Variability 45, 122, 170